Interactive

Differential

Equations

Workbook

 Addison Wesley Interactive

Interactive

Differential

Equations

Workbook

Beverly West
Mathematics
Cornell University

Steven Strogatz
Theoretical and Applied Mechanics
Cornell University

Jean Marie McDill
Mathematics
California Polytechnic State University,
San Luis Obispo

John Cantwell
Mathematics
Saint Louis University

Hubert Hohn
Software Designer
Massachusetts College of Art
Addison Wesley Interactive

Addison Wesley Interactive

Producer: Victoria Zaroff
Project Manager: Kim T. M. Crowley
Production Manager: Lee Stayton
Workbook Production: Cindy M. Johnson and Suzanne Guiod, Publishing Services
Mathematical Workbook Illustrations: Benjamin Halperin
Non-mathematical Workbook Illustrations: Katrina Thomas
Marketing Manager: Liz O'Neil
Cover and Package Design: Dusan Koljensic
Manufacturing Coordinator: Beverly Brissette

ISBN: 0-201-57132-3

1 2 3 4 5 6 7 8 9 10 CRW 00999897

Table of Contents

Part IV   Systems of Differential Equations

Part V   Chaos and Bifurcation

Part VI   Series Solutions and Boundary Value Problems

To the Student

Interactive Differential Equations (IDE) offers a different approach to the exploration of differential equations than you are likely to have encountered before. IDE is based on a collection of interactive illustrations that are very useful for classroom demonstration and discussion, laboratory and homework assignments, or independent study. By visualization and interaction we bring new approaches to the task of conquering new concepts. These illustrations are not intended to replace an open-ended differential equations graphics tool, which can take you as far as you want to go once you know where you are going, but rather to help you build a clear understanding of concepts when you first encounter them, and to help you over the hurdles of abstraction that have often confused our students in the past. In addition, since each tool focuses on one topic, you can get a much deeper understanding of many differential equations topics from the IDE tools than you can from an open-ended differential equations graphics solver. For many who study mathematics, a picture is indeed worth a thousand words (or at least quite a few). Visualization through interactive graphics and animations are especially powerful in differential equations. These interactive graphics tools are designed to help you absorb concepts "into your bones."

You will repeatedly be asked to experiment, ask and/or answer questions, make conjectures, and try to verify them. The labs in IDE will challenge you with unexpected questions. After (or before) you see a graphic representation of a concept, you will often be asked to explain the same phenomenon in algebraic or numerical terms. The variety of approaches usually brings new insights—and new questions—hence, the cycle never ends. We hope you will enjoy, and profit from, the time you spend with IDE.

Beverly West, *Cornell University*, NY
Steven Strogatz, *Cornell University*, NY
Jean Marie McDill, *California Polytechnic State University, San Luis Obispo*, CA
John Cantwell, *Saint Louis University*, MO

Hubert Hohn, *Massachusetts College of Art*, MA

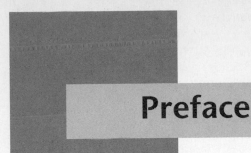

Preface

What is *Interactive Differential Equations* (IDE)?

The reform movement in mathematics places new emphasis on conceptual understanding and active learning. Instead of memorizing formulas or rote solutions, students are encouraged to explore and discover mathematical concepts through writing or illustration. The impact of interactive computer graphics on differential equations has begun a revolution in the teaching of the course. While improvements in visualization through interactive graphics and animations are effective in many disciplines, they are especially powerful in differential equations, where motion and change over time are central issues. Interactive Differential Equations (IDE) brings the strength of this new medium to the learning of differential equations by providing those who teach differential equations a new kind of tool to illustrate abstract examples and encourage students' conceptual understanding.

IDE consists of a CD-ROM containing over 90 interactive tools and a workbook with labs and exercises that relate to those tools. Applications are drawn from engineering, physics, chemistry, and biology. Topics span first- and second-order differential equations, linear and nonlinear systems, Laplace transforms, boundary value problems, and series solutions. By bringing IDE into classroom demonstration and discussion, laboratory and homework assignments, or independent study, teachers and students can explore the mathematics of time and motion with these unique interactive illustrations and corresponding labs.

Students might answer questions by writing, exploring an IDE tool, or drawing illustrations of solution trajectories. Through the visualization and interaction of IDE, new approaches can be used to tackle key concepts.

Here are some distinctive features of IDE:

- IDE is easy to use, with no syntax or special languages to learn.

- Students manipulate physical models, equations, and graphs in "real time."

- The labs from the IDE workbook were written by leading DE educators and include writing exercises, algebraic manipulation, and important links to real-world situations.

- IDE is a bridge to any open-ended differential equations graphics tool or computer algebra system. IDE is designed to build a clear understanding of concepts when they are first encountered by students and to help them over the hurdles of abstraction that have often confused students in the past. An open-ended tool can then build and expand on the topics first introduced with IDE.

- Textbook mapping allows for easy integration into any existing Differential Equations course. When you select the textbook you are using in your course from a list of the major textbooks, IDE will offer appropriate choices of tools and labs for your DE course. See the list of textbooks in the Resources section.

How can IDE be used?

Lecture. IDE can effectively demonstrate concepts that are difficult for educators to illustrate with chalk or with other computer software.

Classroom discussion. Instructors can pose important questions to their students using IDE.

Laboratory. IDE can be set up to be used in a laboratory environment, with students exploring core concepts on their own or with an instructor.

Homework. IDE can be used by students in homework assignments or lab projects. Labs can be broken up into smaller parts for smaller homework assignments.

Independent Study. In combination with a textbook, IDE can bring differential equations to life for students studying on their own.

Other Uses. Visit the IDE Web site at http://awi.aw.com/mathematics/ide/ide.html to understand how people are using IDE in their courses.

IDE is specially tailored to ordinary differential equations courses, but can also be used in courses in non-linear dynamics, chaos, dynamical systems, engineering dynamics, classical mechanics, systems theory, physics, and mathematical biology.

The IDE Workbook

The IDE workbook contains labs and exercises organized by topic area. Each lab starts with a brief introduction to the concept or application covered, and uses one or more IDE tools to explore the concept. The labs are designed to lead students to a deeper understanding of important topics by posing questions that require them to draw graphs, write about the mathematics, or relate to underlying algebraic concepts. Most importantly, the labs provide students a context in which they can use the IDE tools to understand differential equations.

General Tool Instructions

Follow these basic instructions to manipulate each of the IDE tools. Detailed tool instructions for the tools used in each lab can be found at the end of the lab.

Quit Click the mouse on the [**Menu**] button in the upper left corner of the screen to quit the program and return to the contents screen. Do the same to recover from a screen saver attack.

Adjust Parameters
- Use the sliders to set or change a value of a parameter. Press the mouse down on the slider knob for the parameter you want to change and drag the mouse back and forth, or in some cases up and down, or click the mouse in the slider channel at the desired value for the parameter.
- Clicking in a plane while a trajectory is being drawn will start a new trajectory.

Equations
- Click the button to the left of the equation to scroll the list of equations.
- Click an equation to select it.

Buttons
- Click the mouse on the [**Draw Field**] button to draw a slope field.
- Click the mouse on the [**Start**] button to start a trajectory using a preset initial condition.
- Click the mouse on the [**Pause**] button to stop a trajectory without canceling it.
- Click the mouse on the [**Continue**] button to resume the motion of the paused trajectory.
- Click the mouse on the [**Clear**] button to remove all trajectories from the graphs.

Resources

Textbook Mapping IDE is mapped to the following textbooks:

Abell/Braselton, *Modern Differential Equations: Theory, Applications, Technology*
Harcourt Brace & Company, 1996.

Blanchard/Devaney/Hall, *Differential Equations, Preliminary Ed.*
PWS Publishing, 1996.

Borrelli/Coleman, *Differential Equations: A Modeling Perspective, Preliminary Ed.*
John Wiley & Sons, Inc, 1996.

Boyce/DiPrima, *Elementary Differential Equations with Boundary Value Problems, 5th Ed.*
John Wiley & Sons, Inc, 1992.

Bronson, *Schaum's Theory and Problems: Differential Equations, 2nd Ed.*
McGraw-Hill, 1996.

Davis, *Differential Equations for Mathematics, Science, and Engineering*
Prentice Hall, Inc, 1996.

Edwards/Penney, *Differential Equations and Boundary Value Problems: Computing and Modeling*
Prentice-Hall, Inc, 1996.

Edwards/Penney, *2nd Elementary Differential Equations with Boundary Value Problems, 2nd Ed.*
Prentice-Hall, Inc, 1989.

Farlow, *Introduction to Differential Equations and Their Applications*
McGraw-Hill, Inc, 1994.

Guterman/Nitecki, *Differential Equations: A First Course, 3rd Edition*
Harcourt Brace Jovanovich, 1992.

Hubbard/West, *Differential Equations: A Dynamical Systems Approach*
Springer-Verlag, 1995.

Kostelich/Armbruster, *Introductory Differential Equations: From Linearity to Chaos*
Addison-Wesley Publishing Company, 1996.

Kreyszig, *Advanced Engineering Mathematics, 6th Ed.*
John Wiley & Sons, Inc, 1988.

Kreyszig, *Advanced Engineering Mathematics, 7th Ed.*
John Wiley & Sons, Inc, 1992.

Lomen/Lovelock, *Exploring Differential Equations via Graphics and Data, Preliminary Ed.*
John Wiley & Sons, Inc, 1996.

Nagle/Saff, *Fundamental of Differential Equations, 3rd Edition*
Addison-Wesley Publishing Company, 1993.

Nagle/Saff, *Fundamentals of Differential Equations, 4th Edition*
Addison-Wesley Publishing Company, 1996.

Nagle/Saff, *Fundamentals of Differential Equations with Boundary Value Problems, 2nd Ed.*
Addison-Wesley Publishing Company, 1996.

Ross, *Differential Equations, 3rd Ed.*
John Wiley & Sons, 1984.

Ross, *Differential Equations: An Introduction with Mathematica*
Springer-Verlag, 1995.

Simmons/Robertson, *Differential Equations with Applications and Historical Notes, 2nd Ed.*
McGraw-Hill, Inc, 1996.

Strogatz, *Nonlinear Dynamics and Chaos*
Addison-Wesley Publishing Company, 1994.

Zill/Cullen, *Differential Equations with Boundary-Values Problems, 3rd Ed.*
PWS-Kent Publishing Company, 1993.

Zill/Cullen, *Differential Equations with Boundary-Values Problems, 4th Ed.*
PWS-Kent Publishing Company, 1996.

Zill, *A First Course in Differential Equations with Modeling Applications, 6th Ed.*
PWS-Kent Publishing Company, 1996.

Bibliography

Armenti, Angelo Jr., ed. *The Physics of Sports.* (New York: American Institute of Physics, 1992).

CODEE Newsletter, by NSF Consortium on Teaching Ordinary Differential Equations with Computer Experiments. To subscribe, send complete address to CODEE, Math Dept., Harvey Mudd College, Claremont, CA 91711 (codee@hmc.edu).

College Mathematics Journal (MAA) Vol. 25, November 1994 Special Issue on innovation in teaching differential equations, including annotated bibliography of available software.

Countryman, Joan. *Writing to Learn Mathematics* (Heinemann Press, 1992)

Erlichson, Herman. "Maximum Projectile Range with Drag and Lift, with Particular Application to Golf." *American Journal of Physics* 51: 357–362 (1983).

Felsager, Björn and Beverly West. *Laboratory Manual for Differential Equations* (Mathematics Department, Cornell University, 1991)

Gruzka, Thomas. "A Balloon Experiment in the Classroom." *College Mathematics Journal* 25: 442–444 (1994).

Hubbard, McDill, Noonburg, and West. "A New Look at the Airy Equation, with Fences and Funnels." *College Mathematics Journal* 25: 419–43 (1994).

Lanchester, F. W. *Aerodonetics.* (London, 1908).

MacDonald, William M. and S. Hanzely. "The Physics of the Drive in Golf." *American Journal of Physics* 59: 213 (1991), and see the references cited therein.

McDill, J.M. and Björn Felsager. "The Lighter Side of Differential Equations," *College Mathematics Journal* 25: 448–452 (November 1994).

Miele, Angelo. *Flight Mechanics.* Vol. I, *Theory of Flight Paths.* (Reading, MA: Addison-Wesley, 1962).

Minton, Ronald. "A Progression of Projectiles: Examples from Sports." *College Mathematics Journal* 25: 436–442 (1994).

Strogatz, Steven. "Love Affairs and Differential Equations" *Mathematics Magazine* 61: 35 (February 1988).

Strogatz, Steven. *Nonlinear Dynamics and Chaos, with Applications to Physics, Biology, Chemistry and Engineering.* (Reading: Addison-Wesley, 1994).

von Mises, Richard. *Theory of Flight.* (New York: Dover, 1959).

Differential Equations Graphing Software

Differential Systems, Gollwitzer, Herman	Macintosh	(U'Betcha Publications (215) 544-9257)
MacMath 9.2, Hubbard, J.H. and B. H. West	Macintosh	(Springer Verlag, 1995 (800) 777-4643)
Phaser, Koçak, H.	PC compatible	(Springer Verlag 1989 (800) 777-4643)
MDEP, U.S. Naval Academy	PC compatible	(Mathematics Dept., USNA (301) 267-3892)

Computer Algebra Systems

Derive	PC compatible	(Soft Warehouse (808) 734-5801)
Maple	Multiplatform	(Waterloo Maple Software (519) 747-2373)
Mathematica	Multiplatform	(Wolfram Research (217) 398-0700)
MatLab	Multiplatform	(Mathworks (508) 653-1415)

Author Acknowledgments

For help in making the development of IDE possible, we wish to thank:

Meg Hickey, who suggested some of the innovative ideas for Laplace transforms, worked the problems in all the labs, and provided numerous suggestions for improving the text and software.

Mike Klucznik, for a close reading and much useful discussion of the topics.

Cindy Johnson and Donovan Hohn, who provided typesetting, thoughtful text editing, and nice solutions to lots of difficult matters of presentation.

Paul Lindale, for designing a lovely interface to accommodate our textbook mapping.

Ben Halperin, for mathematical illustrations, and Katrina Thomas, for the other illustrations.

Kim Crowley, our project manager, who pulled it all together and kept us all going.

Vicky Zaroff and Chip Price at Addison Wesley Interactive, who gave us the chance to work together on the development of these materials.

We give extra thanks to Vicky and Kim for jumping into the trenches with us at key moments, thus assuring both that the project would get done, and that it could accomplish so much.

Reviewers of IDE

Thank you to all the alpha reviewers...

Roy Alston, *Stephen F. Austin State University*

Joel Anderson, *Pennsylvania State University*

Ararat Andrasian, *Montgomery College*

Beverly Baartmans, *Michigan Technological University*

Richard Burns, *Springfield Technical Community College*

John Davenport, *Georgia Southern University*

Gary Epstein, *California Polytechnic University, San Luis Obispo*

Benny Evans, *Oklahoma State University*

Karla Foss, *Pellissippi State Technical Community College*

Constant Goutziers, *State University New York, Oneonta*

Bruno Guerrieri, *Florida A&M University*

D. J. Herbert, *University of Pittsburgh*

Lorraine Holub, *Virginia Polytechnical Institute*

Sergio Loch, *Grand View College*

Christine McMillian, *Virginia Polytechnical Institute*

Gordon Melrose, *Old Dominion University*

Lawrence Moore, *Duke University*

Timothy Murdoch, *Washington & Lee University*

Kristin Pfabe, *Northern Kentucky University*

Zwee Reznik, *Fresno City College*

Lila Roberts, *Georgia Southern University*

Ashok Sen, *Purdue University*

Eileen Shugart, *Virginia Polytechnical Institute*

Robert Wheeler, *Virginia Polytechnical Institute*

And all the beta reviewers...

Carol Adjemian, *Pepperdine University*

Bill Albrecht, *Pasco-Hernando Community College*

Richard Barshinger, *Pennsylvania State University, Scranton*

Paul Blanchard, *Boston University*

Richard Bedient, *Hamilton College*

Erik Bollt, *United States Military Academy*

Jerald Dauer, *University of Tennessee*

Matt Dempsey, *Jackson Community College*

George Dorner, *William Rainey Harper College*

Stephen Drake, *Northwestern Michigan College*

Robert Garry, *Hawaii Community College*

G.S. Gill, *Brigham Young University*

Thomas Gruszka, *Western New Mexico University*

Richard Hall, *Boston University*

Gene Hamilton, *Washington College*

Donald Hartig, *California Polytechnic State University, San Luis Obispo*

William Harris, *University of Southern California*

Silvia Heubach, *California State University, Los Angeles*

Aimee S. A. Johnson, *Swarthmore College*

David Kerr, *Eckerd College*

Allen Killpatrick, *University of Redlands*

Aaron Klebanoff, *Rose-Hulman Institute of Technology*

Carl Main, *Shoreline Community College*

Joe Marlin, *North Carolina State University*

Mike Martin, *Johnson Community College*

Elly Claus-McGahan, *University of Puget Sound*

Robert McOwen, *Northeastern University*

Ruth Michler, *University of North Texas*

Rennie Mirollo, *Boston College*

Darryl Nester, *Bluffton College*

Martin Peres, *Broward Community College*

Rudra Pratap, *Cornell University*

John Ringland, *State University of New York, Buffalo*

James Schlesinger, *Tufts University*

Alexandra Skidmore, *Emory and Henry College*

Robert Snyder, *Cornell University and Simon's Rock of Bard College*

John Sylvester, *University of Washington*

Rebekah Valdivia, *Augsburg College*

Rick Vitray, *Rollins College*

Weiqing Xie, *California Polytechnic University, Pomona*

Phil Zenor, *Auburn University*

About the Authors

Beverly West
Beverly West of Cornell University, author of several leading books on differential equations, has been an active member of C-ODE-E, a National Science Foundation-sponsored team to promote graphics experimentation in differential equations courses.

Steven Strogatz
A recipient of numerous awards, including MIT's highest teaching prize and a Presidential Young Investigator Award from the National Science Foundation, Steven Strogatz is an expert on nonlinear dynamical systems in engineering, physics, and biology. He is now in the Theoretical and Applied Mechanics Department at Cornell University.

Jean Marie McDill
As an instructor at a university with a large undergraduate engineering program, Jean Marie McDill of California Polytechnic State University, San Luis Obispo, has experimented widely with the uses of technology in the introductory course in differential equations. She now includes a laboratory component in the differential equations course.

John Cantwell
John Cantwell of Saint Louis University, an expert in geometric topology and foliations, uses visualization to help his students understand differential equations concepts. He is a veteran teacher who has long been interested in the use of technology in teaching.

About the Software Designer

Hubert Hohn
Hubert Hohn designed and developed the interactive tools found in IDE. He is the director of the Computer Arts Center of the Massachusetts College of Art and specializes in developing interactive visual tools for mathematics education.

About the Publisher

Addison Wesley Interactive (AWI) is a publisher of interactive media products for college-level courses in mathematics, engineering, physics, and statistics. AWI is a business unit within the Addison Wesley Longman Publishing Company.

The team of high-tech professionals and software designers at AWI works with teams of leading educators to design our products. The result of this collaboration among publisher, educators, and multimedia engineers is a new generation of interactive learning tools. Working with AWI products, students can visualize, relate to, and ultimately master complex concepts more quickly and more thoroughly than ever before.

Interactive Differential Equations (IDE) is our first product developed for the mathematics curriculum. Its emphasis on multiple representation and active learning provides instructors with a new teaching and learning resource for the differential equations classroom.

To find out more about this and other AWI projects, please call Liz O'Neil at 617.944.3700 ext. 2380, e-mail awi-info@aw.com, or visit our Web site at http://awi.aw.com/.

Part I

First Order Differential Equations

Newton's Law of Cooling

1

A cup of coffee cooling on the counter, a cake warming in the oven, and a body found in the chill autumn weather . . . are these the ingredients for a murder mystery to read by the fire or a case for that most famous detective of natural phenomena, Sir Isaac Newton? We use a first-order linear differential equation formulated by Newton to predict the temperatures of objects introduced into media with known ambient temperatures.

1. A Basic Differential Equation

A simple method to model the cooling or heating of an object placed in a constant ambient temperature is to say that the time rate of change in temperature is proportional to the difference between the temperature A of the surrounding medium (the ambient temperature) and the temperature T of the object:

$$\frac{dT}{dt} = k\big(A - T(t)\big) \quad \text{for } k > 0. \tag{1}$$

This equation is called **Newton's Law of Cooling and Heating.** In this model Newton assumed that the heat transfer between the object and the surrounding environment is not sufficient to affect noticeably the ambient temperature. To investigate this model, open the **Newton's Law of Cooling: Curve Fitting** tool. For convenience, we will denote the time rate of change in temperature by $\dot{T} \equiv \frac{dT}{dt}$.

1.1 If $T > A$, is the object cooling or heating? Is $\frac{dT}{dt}$ positive, negative, or zero when $T > A$?

1.2 For what value of T is $\frac{dT}{dt} = 0$? This value is called the **equilibrium** value.

3

1.3   If $T(0) = T_0$, solve equation (1) analytically to show that $T = A + (T_0 - A)e^{-kt}$.

2. Graphical Representations

Graphical information can be displayed in several formats. To familiarize yourself with some types of graphical displays, open the **Newton's Law of Cooling: Cooling Rate** tool. To set an initial temperature, click the mouse on a point in the upper-left graphical window. Note that the other graphical windows also respond to this initial setting. Experiment with various values for the constant of proportionality k and the ambient temperature A.

2.1   What do you notice about the behavior of the curve on the $\dot{T}$ vs. t graph (in the upper-right

corner)? What happens to $\dot{T}$ as t becomes large? Vary k and

A on the sliders. Is the long-term   behavior of $\dot{T}$ always the same?

2.2   Look at the graph of $A - T$ vs. $\dot{T}$. How do you interpret the straight-line graph? What
does the slope of the line denote?

3. Real-World Connections

3.1   Open the **Newton's Law of Cooling: Curve Fitting** tool and select the **[Show Coffee Data]** button.
Data collected from H. Hohn's cup of coffee are displayed on the graph as data points. By carefully
fitting the curve to the data, determine the appropriate values for A, k, and T_0. Record them below.

$A =$
$k =$
$T_0=$

3.2   Use the solution of Equation (1) that you found in Exercise **1.3** to determine the time when the
coffee is 180 degrees. Is this time consistent with your graph?

The general solution for Equation (1) is $T = A + Ce^{-kt}$. If the ambient temperature A is given, then two data
points are required to determine the constants C and k.

3.3   Coroners use several methods to determine time of death. If Equation (1) were used, measurements of the temperature at two different times would be required to establish k and the constant of integration. Suppose this were the only method used to determine time of death in a case where the time of death was the crucial element in the prosecution's case. How would you, as the scientific consultant, help the defense cast doubt on this estimate? Think carefully about the assumptions of the model!

Lab 1: Tool Instructions

Newton's Law of Cooling: Curve Fitting Tool

Parameter Sliders

Use the slider to change the values for the parameters T_0, A, and k.

Press the mouse down on the slider knob for the parameter you want to change and drag the mouse up and down, or click the mouse in the slider channel at the desired value for the parameter.

Buttons

Click the mouse on the **[Show Coffee Data]** button to show the coffee data on the graph.
Click the mouse on the **[Hide Coffee Data]** button to hide the coffee data.

Newton's Law of Cooling: Cooling Rate Tool

Setting Initial Conditions

Click the mouse on the tT graphing plane on the top left (labeled T) to set the initial conditions for a trajectory.
Clicking in the plane while a trajectory is being drawn will stop the trajectory and start a new one.

Parameter Sliders

Use the sliders to change the values for the parameters A and k.

Press the mouse down on the slider knob for the parameter you want to change and drag the mouse left or right, or click the mouse in the slider channel at the desired value for the parameter.

Buttons

Click the mouse on the **[Clear]** button to remove all trajectories from the graph.

Graphing Differential Equations

2

Tools Used in Lab 2
Slope Fields
Solutions

A first-order differential equation is an equation for which the highest-order derivative is first-order. The goal is to find the function, knowing its derivative. How can we do this with graphs?

Introduction

A first-order differential equation has a whole family of solution curves, like a pot of spaghetti with waves of nonintersecting strands. With the tools for this lab, you can investigate such families, and also examine individual members of a family of solutions.

We purposely use t as the independent variable, because it is often helpful (and usually appropriate in context) to think of solutions evolving in time.

1. The Slope Field (or Direction Field)

A first-order differential equation, $\frac{dx}{dt} = f(t, x)$, gives us, for any point (t,x) on the tx-plane, the **slope** of a solution curve. Just as a flag reveals the particular direction of the air current at a flagpole, $\frac{dx}{dt}$ reveals the direction (slope) of a solution curve $x = g(t)$ at any point (t,x) we choose. An aerial view of flags on a grid of flagpoles in a field would show the overall pattern of different wind currents in the field. Similarly, vectors with the slope $\frac{\Delta x}{\Delta t}$, on a grid of points on the tx-plane, indicate the flow of solution curves. Although $\frac{dx}{dt} = f(t, x)$ does not tell us directly the solutions $x = f(t)$, it does give the *slope* of the solution curve at any point. We know from differential calculus that the difference quotient $\frac{dx}{dt}$ gives a tangent line approximation of the curve $x = g(t)$ at every point on the curve. Our aim is to use $\frac{\Delta x}{\Delta t}$ as an approximation to $\frac{dx}{dt}$ to see the flow pattern of the solution curves.

1.1   With the **Slope Fields** tool you can see at any point (t,x) a vector with slope $\dfrac{dx}{dt}$. The slope calcula-

tion is shown in a separate window. By clicking the mouse, you can plot a segment of the slope

vector proportional to the size of the time step. Note how the slope at each point is calculated from

the formula for $\dot{x}$. All vectors indicate the direction of advancing time, t.

Choose one of the given equations to study, and make a picture of the slope field by setting down
enough vectors to give an idea of what will happen wherever you might choose a starting point
(t,x). Sketch the result here:

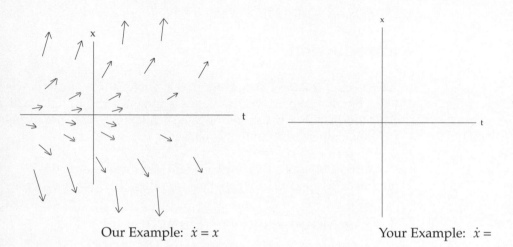

Our Example: $\dot{x} = x$ Your Example: $\dot{x} =$

1.2   Now the object is to find out what happens to a point set in this slope field. What path will be
followed as a point is carried along by the flow, especially in the long run? The focus of studying
differential equations is to be able to predict what will happen, given sufficient starting information.

To *see* a solution—the graph of the function that follows the flow—choose an initial point (t,x) on the
slope field and click the mouse there to plot an arrow. At the head of that arrow, start another one,
and so on, to build a crude approximation to a solution, $x = g(t)$. Your choice of an **initial condition**
(t_0, x_0) picks out a unique member of the family of solutions.

You can click on the **[Draw Field]** button to get a whole grid of direction lines. Notice how your
approximate solution follows all the little lines in the slope field. Fill in a drawing of an approximate
solution on the slope field you have drawn in Exercise **1.1**.

1.3   Try changing the choice of the time step, Δt. What happens to the size of the vector when you
change from $\Delta t = 0.5$ to 1.0?

                    to 0.1?

                    to 0.01?

Why do you end up with nothing but an arrowhead in the last case?

1.4   Now you are ready to let the computer do more of the work. If you click on the **[Solutions]** button, under **Drawing Mode**, every click on the slope field draws from that point an approximate solution computed with a small time step, Δt. By choosing more initial conditions, you can cover the tx-plane with as many solutions as you like to show what behaviors can be expected of the solutions for your equation, in the region defined by the graph. Thus you create a picture of the whole *family* of solutions to this particular differential equation. Sketch here a sampling of the solutions to your example.

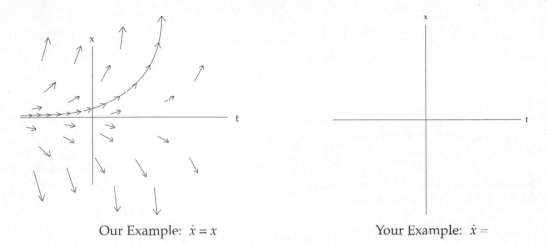

Our Example: $\dot{x} = x$ Your Example: $\dot{x} =$

1.5   Write a verbal description of the behaviors of the solutions for your example, answering the following questions:
 • How do these graphical solutions differ? by a constant? otherwise?
 • Are there horizontal translates? vertical translates?
 • Do they merge? do they diverge? where, and how?
 • Do any solutions appear to have vertical asymptotes? horizontal asymptotes?
 • Do any solutions appear to be **equilibria** (to have a constant value of x)? Do they connect to any of your answers above?

1.6   Look at a few more of the equations in the **Slope Fields** tool with the preceding questions in mind. Show one as an example, with your comments. Then, if you have not looked at $\dot{x} = \cos x$, try it now. Predict where the equilibrium solutions are.

$\dot{x} =$ $\dot{x} = \cos x$

2. Introduction to an Open-Ended Differential Equations Graphing Tool

An open-ended tool, which could be
- a graphing calculator
- a dedicated interactive differential equations program, such as
  MacMath, Differential Systems, MDEP, Phaser*
- a computer algebra system such as *Derive, Maple, Mathematica*, or *Matlab**

lets you enter a function of your own choice for the differential equation, and change the bounds on the window. It also allows you to change the numerical approximation method and/or the stepsize to see the results.

Try whatever open-ended tool is at hand on one of the differential equations already studied (or on something different if you choose), to learn what it can do. Then you will henceforth have at your disposal a means to explore examples beyond those we have selected for the labs.

Make a print (or sketch) for your example, to show the effects of two different stepsizes, with Euler's method, on the same set of initial conditions. Annotate your printout with a little discussion of the results and how you think they could be explained.

3. Comparison with Algebraic Solutions

You have probably noticed that we have used no algebra, no calculus, and no obvious numerics up to this point, but rather have concentrated on qualitative pictures and ideas.

The actual way in which the solutions are drawn on the tx-plane is by numerical calculation of very short vectors using the slopes and small time steps. This is discussed in some detail in Lab 5, Numerical Methods.

3.1 We now turn our attention to those cases (a minority, except in differential equations textbooks) in which it is possible to find an algebraic formula for solutions by such methods as separation of variables or integrating factors for linear equations. This is called an **analytical** solution. As an example, we consider $x' = x - t$, for which the solution is $x = t + 1 + Ce^t$. Confirm that this is in fact the solution, either by analytical methods, or simply by differentiating to show that it indeed satisfies the differential equation:

3.2 Using the **Solutions** tool, choose an equation and start a solution; the computer then automatically draws both an approximate solution and the analytical solution (in different colors) through the same initial condition, after calculating the proper value of the constant C. Usually these curves look similar, but not exactly the same. How can you show that they differ? The kinds of differences you see depend on the example you have chosen.

*For bibliographic details on any of the tools listed, see the list of resources in the front matter of this book.

First for $\dot{x} = x - t$, then for $\dot{x} = x \cos t$, set stepsize to 0.5 and method to Euler. Make a two-colored sketch to show what differences appear. For instance, for a given initial condition, are the intercepts at different coordinates on the approximate and analytical solutions? Or do they hit the edges of the graphs at different values? Do their vertical distances grow consistently?

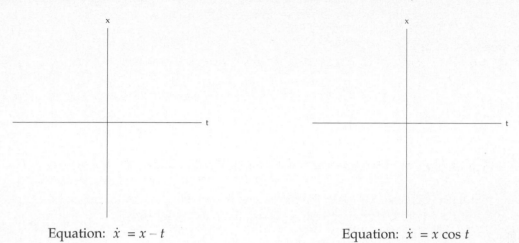

Equation: $\dot{x} = x - t$ Equation: $\dot{x} = x \cos t$

Note: You may see numerical approximations become downright *wrong* when the timestep is too big for the time axis. A large time step occasionally causes sharp zigzags that look, and are, obviously, not correct. Such "jaggies" are illustrated in Lab 6, Isoclines and Fences, Exercise **2.5**.

3.3   For either $\dot{x} = x + t$ or $\dot{x} = x \cos t$, make a table listing several initial conditions and whatever measure seems appropriate to illuminate the difference(s), for example, t or x intercepts for approximate and algebraic solutions, or final values for t or x, or maximum vertical distance between approximate and algebraic solutions. Add whatever else you might wish to compare, such as values of the constant C, or the long term behavior of solutions.

your equation: initial conditions		comparison: intercepts or _____			
t_0	x_0	approximate	analytical		

3.4   Write a paragraph about what you can observe from this experiment. Where is the approximate solution close to the analytical solution? Where is it not close?

As you study differential equations, you should become able to predict when the approximate numerical solution, computed in time steps, will not be a good fit to the analytical solution, computed algebraically.

3.5   With the solutions tool on these equations, try smaller stepsizes, and try Runge-Kutta approximations. Discuss the sort of improvements you seem to find for each.

Smaller stepsizes:

Runge-Kutta:

You will further explore these improvements in Lab 5, Numerical Methods.

3.6   Following is the list of the equations in the **Slope Fields** tool. For each of them, find the analytical (algebraic) solution where possible; otherwise, mark them as "not possible."

1) $\dot{x} = x$

2) $\dot{x} = \cos t$

3) $\dot{x} = \cos x$

Does your solution to 3) include the equilibrium solutions found in Exercise **1.6**? These must often be found separately.

4) $\dot{x} = x \cos t$

5) $\dot{x} = \cos tx$

6) $\dot{x} = x + t$

7) $\dot{x} = x - t$

8) $\dot{x} = x^2 - t$

9) $\dot{x} = x^2 - t^2$

10) $\dot{x} = x^2 + t^2$

11) $\dot{x} = \cos(x^2 + t^2)$

12) $\dot{x} = -tx$

For each of these functions that has an analytical solution, the graphs of the numerical and analytical solutions can be compared with the **Solutions** tool. For those that do not, we have to rely on the graphical pictures of the solutions, but studying the other equations may start to give you some intuition about the accuracy of the numerical solutions in those cases as well.

3.7 For the equations in Exercises **3.6** where you know an analytical solution, explain how the formula confirms your visual expectation or forces you to revise some of the answers you made in Section 1. Comparison of analytical and numerical solutions can be used to advance your understanding of mathematics—so check carefully how the formulas compare with your descriptions, and acknowledge where refinements can now be made. Use a separate page if necessary.

4. Follow-Up Exercises for Classwork

Examine the twelve slope fields, with graphical solutions, as studied in the **Slope Fields** tool. Pictures are provided on the next two pages.

4.1 Solution curves that have identical shapes but different vertical and/or horizontal positions are called **translates**. Which of these slope fields have some solutions that are

horizontal translates?

vertical translates?

4.2 Which of these slope fields have

solutions with vertical asymptotes?

solutions with horizontal asymptotes?

solutions with oblique asymptotes?

merging solutions?

diverging solutions?

4.3 Can you make some conjecture(s) from the pictures about how to predict any of the above?

5. Slope Field Graphs

Pictures are provided of the twelve slope fields for equations in the **Slope Fields** tool. Here we use a larger "window" ($-10 \leq t \leq 10, -8 \leq x \leq 8$) than in the tool ($-2 \leq t \leq 2$, $-2 \leq x \leq 2$), in order to show more of the overall pattern.

1) $\dot{x} = x$

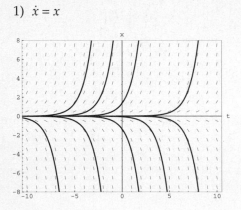

2) $\dot{x} = \cos t$

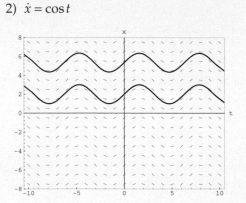

3) $\dot{x} = \cos x$

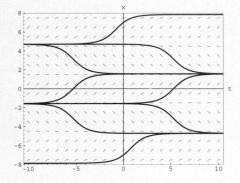

4) $\dot{x} = x \cos t$

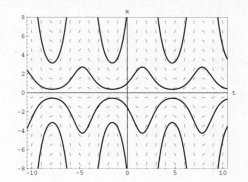

5) $\dot{x} = \cos tx$

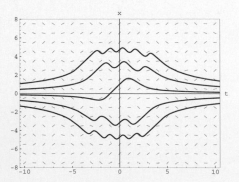

6) $\dot{x} = x + t$

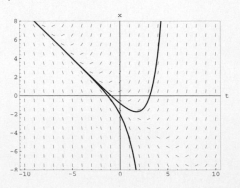

7) $\dot{x} = x - t$

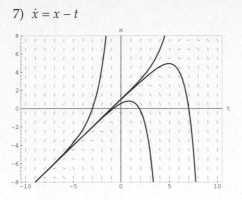

8) $\dot{x} = x^2 - t$

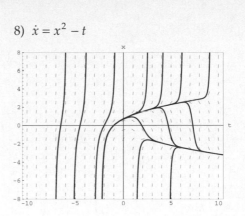

9) $\dot{x} = x^2 - t^2$

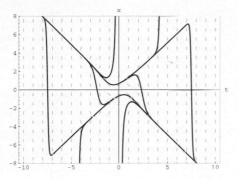

10) $\dot{x} = x^2 + t^2$

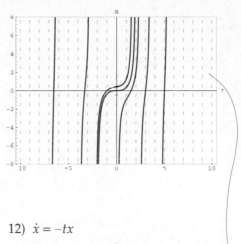

11) $\dot{x} = \cos(x^2 + t^2)$

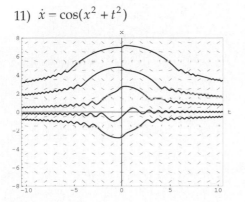

12) $\dot{x} = -tx$

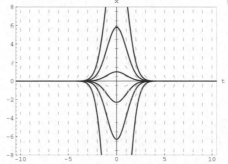

6. Additional Exercises

6.1   Use an open-ended differential equations graphing tool to make an analysis similar to that in Sections 1, 3, and 4 for the following equations:

a. $\dot{x} = -\dfrac{1}{x^2}$ b. $\dot{x} = x^3$

6.2   Based on the similar equations in the **Slope Fields** tool, predict the behavior of the solutions for the following equations. Then use your open-ended differential equations graphing tool to make pictures of the solutions to at least one of the following equations. On a separate page, make sketches of the results, and add annotations that explain where your predictions were fulfilled, and where they were not. In the latter case, note what it was that made the picture differ from your expectations.

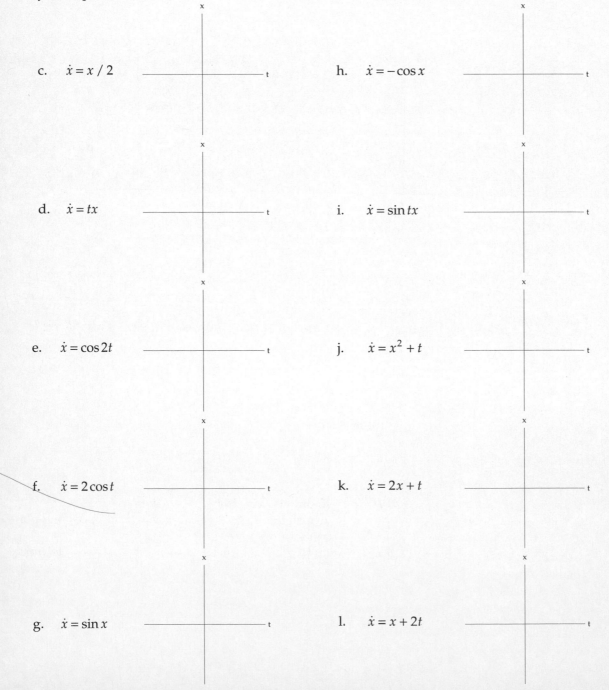

c. $\dot{x} = x/2$ h. $\dot{x} = -\cos x$

d. $\dot{x} = tx$ i. $\dot{x} = \sin tx$

e. $\dot{x} = \cos 2t$ j. $\dot{x} = x^2 + t$

f. $\dot{x} = 2\cos t$ k. $\dot{x} = 2x + t$

g. $\dot{x} = \sin x$ l. $\dot{x} = x + 2t$

Lab 2: Tool Instructions

Slope Fields Tool

Setting Initial Conditions

Click the mouse on the graphing plane to set the initial conditions for a trajectory or a point for a vector.

Clicking while a trajectory is being drawn will stop the trajectory.

Equations

Click the arrow button to the left of the equation to pop up the list of equations.

Click an equation to select it.

Drawing Mode Buttons

Click the mouse on the [Vectors] button to set vectors when you click on the plane.

Click the mouse on the [Solutions] button to display a solution curve when you click on the plane.

Time Step Buttons

Click the mouse on a button in the Δt list to set the time step for vectors and trajectories.

Other Buttons

Click the mouse on the [Draw Field] button to draw a slope field over the graphing plane.

Click the mouse on the [Clear] button to remove all vectors and trajectories from the graphing plane.

Solutions Tool

Setting Initial Conditions

Click the mouse on the tx graphing plane to set the initial conditions for trajectory and define the constant for the solution function.

Clicking while a trajectory is being drawn will stop the trajectory.

When you pass the mouse over the tx plane the functional relationship between the variables is shown.

Equations

Click the arrow button to the left of the equation to pop up the list of equations.

Click an equation to select it.

Time Step Buttons

Click the mouse on a button in the Δt list to set the time step for the trajectories.

Other Buttons

Click the mouse on the [Draw Field] button to draw a slope field over the graphing plane.

Click the mouse on the [Clear] button to remove all slopes and trajectories from the graphing plane.

Click the mouse on the [Euler] button to draw a solution by Euler's method.

Click the mouse on the [Runge Kutta 4] button to draw a solution using the Runge Kutta technique.

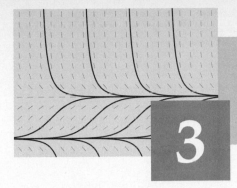

Single Species Population Models

Is there any way to predict how crowded our planet will be in the future? Population growth is often modeled by first-order differential equations. In this lab, we consider only autonomous differential equations—those without explicit time dependence.

1. The Exponential Model

The simplest model for growth and decay is the exponential model

$$\frac{dN}{dt} = rN \, , \qquad\qquad (1)$$

where r is a rate constant and N could be the size of a population, the density of bacteria in a nutrient medium, or a quantity of radium or carbon 14 in a particular state of radioactive decay. This model has solutions that are exponential functions.

1.1 Show the steps involved in finding the solution function $N = N_0 e^{rt}$ for Equation (1), where $N_0 = N(0)$, the initial size of the population.

Open the **Growth and Decay** tool and try some negative values for r. The graph on the left is drawn numerically, using the differential equation, and the graph on the right is drawn using the solution function.

The time required for the amount of decaying material to be reduced by half is called the **half life** and is indicated in red along the solution curve. The time required for a quantity of carbon 14 to halve is about 5,568 years.

1.2 Write a formula for finding the half life. What is the value of r for carbon 14 decay?

For a population model, $N(t)$ is the number of individuals in the population at time t and r is a net growth rate, or birthrate minus death rate. Note that the solution to this differential equation is $N = N_0 e^{rt}$, so that for positive r the population grows without limit. For relatively small populations in a habitat with abundant resources, growth is exponential. The time required for a population to double is an important measure of growth. In the **Growth and Decay** tool, the doubling time is indicated in red alongside the solution function for positive values of r.

1.3   Write the formula for the doubling time of a population growing by the exponential model.

The Breakdown of the Exponential Model

In any real-world situation, totally unrestricted growth is clearly impossible. There are many limiting factors, not the least of which is the fact that the number of atoms in the solar system is finite.

2. The Logistic Equation

The logistic model is based on the exponential growth and decay model, but it includes an overcrowding term, or nonconstant growth rate, that reflects the limitations on growth due to the scarcity of resources and living space. The new term is proportional to the square of the population, so the equation becomes

$$\frac{dN}{dt} = rN - \frac{r}{K}N^2 \qquad \text{or} \qquad \dot{N} = r\left(1 - \frac{N}{K}\right)N \qquad\qquad r, K > 0. \qquad\qquad (2)$$

K is the steady state, or **carrying capacity**, for the population in a particular habitat. The term $\frac{r}{K}N^2$ is a mortality term in which N^2 represents competitive encounters between members of the population. Assuming positive values for N, we can also say that the impact of the growth rate r is diminished by $\left(1 - \frac{N}{K}\right)$ in proportion to the population size. Use the sliders in the **Logistic Growth** tool to observe the changes in the shapes of solutions for various values of the constants r, K, and N_0.

2.1   Describe the changes in population, if any, with increasing time for each of the following three relationships between the population density, N, and the carrying capacity, K:

$N < K$

$N = K$

$N > K$

2.2   Analyze the logistic equation as follows:

a.   Find the equilibrium solutions for the logistic equation and justify this algebraically.

b.   Explain why K is called the **carrying capacity** of the environment.

c. What method (or methods) can you use to solve the logistic equation analytically?

3. The Phase Line

Open the **Logistic Phase Line** tool. The characteristic behavior of the logistic equation is displayed using $K = 1$ and $r = 1$. If x represents the population N, Equation (2) becomes

$$\dot{x} = x(1 - x). \qquad\qquad (3)$$

Up to this point we have been plotting the tx graph and looking at the slope field defined by the slope $\dot{x}$ as a function of time. However, if we project the tx graph onto the x-axis (the vertical axis), we obtain the **phase line**, a succinct graphical summary of the behaviors of solutions. On the phase line, stable equilibrium points are shown as solid dots, unstable equilibrium points are shown as hollow dots, and the directions of flow are designated by arrows. Equilibrium points are also called **fixed points** or **steady states**, and are often designated symbolically with an asterisk, as in x^*. The phase line can also be viewed as a projection of the $x\dot{x}$ graph onto the horizontal axis. The $x\dot{x}$ graph is shown in the upper part of the screen. Clearly $\dot{x}$ as a function of x gives us a parabola that has zeros at $x = 0$ and $x = 1$, corresponding to the fixed points on the phase line.

The phase line is a way to view the **flow.** Think of a fluid flowing along the phase line with a velocity $\dot{x}$ that varies with the value of x according to $\dot{x} = x(1 - x)$, and imagine x as a particle swept along by the flow. As shown on the phase line, the flow is up where $\dot{x}$ is positive, down where $\dot{x}$ is negative, and there is no flow at the fixed points where $\dot{x} = 0$.

3.1 Sketch the graph of solutions to the logistic equation and give a general explanation in words of where stable and unstable equilibria are located. Then show how these are marked on the phase line.

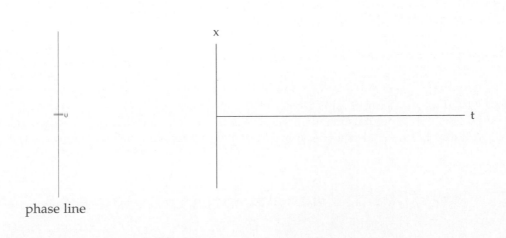

phase line

The **Logistic Phase Line** tool has shown one example of a phase line. In the following exercise, you will construct (by hand) phase lines for other equations.

3.2   For the following autonomous differential equations, use the given information (graph or equation) to locate on the phase line the equilibrium points (with a closed circle for a stable equilibrium and an open circle for an unstable equilibrium) and the appropriate arrows (according to whether x is increasing or decreasing in that region).

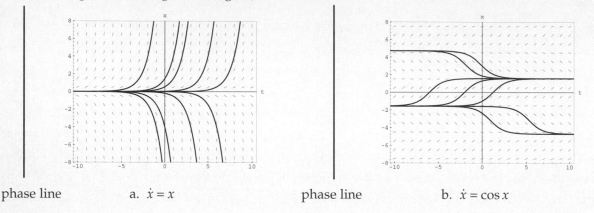

phase line   a. $\dot{x} = x$ phase line   b. $\dot{x} = \cos x$

The following, for which we have not given a picture, should be done simply by analyzing the signs that can appear in $\dot{x}$.

c. $\dot{x} = (x-1)(x-2)(x-3)$ $\dot{x} = \sin\left[(x - \pi/2)x\right]$

phase line   phase line

4. Logistic Growth with Harvesting

We can modify the logistic equation Equation (2) by including a harvesting term $h(t,N)$:

$$\dot{N} = rN(1 - \frac{N}{K}) - h(t,N) \tag{4}$$

For example, if our population were fish, $h(t,N)$ would represent a reduction in numbers due to fishing. Although seasonal variations in fishing might include an explicit dependence on t, we use the autonomous case in this lab. Another reasonable possibility is that the fishing rate might depend on the abundance of fish N. Using the simplified model Equation (3) with a constant harvesting rate h, we get

$$\dot{x} = x(1 - x) - h. \tag{5}$$

Using the **Logistics with Harvest** tool, observe the changes in the behaviors of trajectories when the value of h is varied.

4.1 Constant Harvesting Rate

a. Set h to 0.2. Look at the tx graph. Try several initial values for the population x by moving the cursor up and down on the phase line slider, or across the $x\dot{x}$ graph, to see how the initial population determines the future growth and survival of the population. When is there danger of extinction? Discuss the possibilities. Are there equilibrium populations? If so, what are they?

b. Set $h = 0.4$. Answer the preceding questions again. Are there any equilibrium populations? Can you find any initial population that does not result in extinction?

4.2 Critical Harvesting Rate

a. Experiment with h, starting with $h = 0.4$. Decrease h and look at the graph as well as the phase line. Note that the changes in h affect the number of fixed points and their stability.

Choose h values that give qualitative differences in behavior, and draw the fixed points on the sketches of the phase line. Indicate whether they are stable, unstable, or semi-stable (denoted by a half-filled circle, a semi-stable point attracts on one side and repels on the other). If there are no fixed points, say so! Use arrows to show the flow directions along the phase line, toward or away from any fixed points.

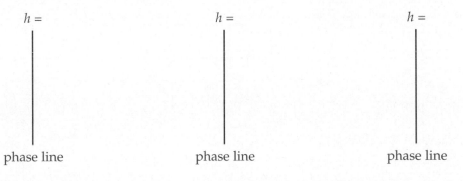

$h =$ $h =$ $h =$

phase line   phase line   phase line

b. Describe what the number and value of fixed points signify about fish populations, over-harvesting, and survival vs. extinction in terms of our model.

i. two fixed points

ii. one fixed point

iii. no fixed points

c. You have just found the critical **harvesting rate**, h_c, the harvesting rate for which there is only one fixed point. The change in the character of the system's behavior across this parameter value is called a **bifurcation**. What is your experimental result for h_c in the simplified model described by Equation (5)?

d. Now consider the general case, Equation (4). To find equilibrium levels (or fixed points) N_1^* and N_2^* you must set $\dfrac{dN}{dt} = 0$ and solve the resulting quadratic equation in N, using a constant h. Show your work.

e. Look at the discriminant using the quadratic formula. Show that N_1 and N_2 are positive if $h < rK/4$ and that there is **exactly one** equilibrium level when $h = rK/4$. Consequently this value of h must be the critical harvesting rate, h_c. Compare this value to your experimental value for h_c (when $r = 1$ and $K = 1$). What is the inescapable result when $h > h_c$?

4.3 Critique of the Model

a. What is wrong with the model? When does it not make sense?

b. Is the following model from Strogatz [SS, p. 90] more reasonable? Justify your answer.

$$\frac{dN}{dt} = rN(1 - \frac{N}{K}) - h\frac{N}{A+N} \quad \text{where } h > 0 \text{ and } A > 0.$$

c. What happens when N gets large? When N gets small?

Lab 3: Tool Instructions

Growth and Decay Tool

Setting Initial Conditions

Click the mouse on the left graphing plane to set the initial conditions for a trajectory and the constant for the solution function.

Clicking while a trajectory is being drawn will stop the trajectory.

When you pass the mouse over the right plane, the functional relationship between the two variables is shown.

Parameter Slider

Use the slider to set the rate constant r.

Press the mouse down on the slider knob for the parameter and drag the mouse back and forth, or click the mouse on the slider channel at the desired value for the parameter.

Buttons

Click the mouse on the **[Clear]** button to remove all trajectories from the graphing planes.

Click the mouse on the **[Drawing Field]** button to draw a slope field over the left graphing plane.

Logistic Growth Tool

Parameter Sliders

Use the sliders to change the values for the constants r, K, and N_0.

Press the mouse down on the slider knob for the parameter you want to change, and drag the mouse up and down, or click the mouse in the slider channel at the desired value for the parameter.

Logistic Phase Line Tool

Setting Initial Conditions

Click the mouse on the tx plane, or the $x\dot{x}$ plane or the phase line to set the initial condition for a trajectory.

Clicking while a trajectory is being drawn will stop the trajectory.

Buttons

Click the mouse on the **[Clear]** button to remove all trajectories from the graphing planes.

Logistic with Harvest Tool

Setting Initial Conditions

Click the mouse on the tx plane, or the $x\dot{x}$ plane or the phase line to set the initial condition for a trajectory.

Clicking while a trajectory is being drawn will stop the trajectory.

Parameter Slider

Use the slider to set the harvest constant h.

Press the mouse down on the slider knob for the parameter and drag the mouse back and forth, or click the mouse in the slider channel at the desired value for the parameter.

Buttons

Click the mouse on the **[Clear]** button to remove all trajectories from the graphing planes.

Mechanics: Falling Bodies and Golf

4

Suppose you're trying to hit a golf ball as far as possible. Should you launch it at a 45-degree angle? Remember, in the real world, you can't always neglect air resistance.

1. Bead Dropping Through Shampoo

In a memorable TV commercial that aired several years ago, a small bead was placed inside a transparent bottle of Prell® shampoo, and allowed to drop ever so slowly through the thick green liquid. The commercial was visually striking because the descent of the bead was so smooth, so gradual, almost hypnotic.

The bead dropping through the shampoo is an example of a body moving through a resistive medium. To take another example, we are all aware of air resistance—you can feel it by sticking your hand out the window of a moving car. Later in this lab we are going to examine the effects of air resistance on the flight path of a golf ball. But first, let's think a bit more carefully about the physics of the shampoo problem.

The motion of the bead is governed by Newton's law $F = ma$. If $v(t)$ denotes the bead's velocity (measured positive downward, for convenience), then Newton's law becomes

$$m\frac{dv}{dt} = mg - bv$$

where m is the mass of the bead, g is the acceleration due to gravity, and b is a measure of the viscosity, or frictional resistance, provided by the shampoo. Here, the drag force $-bv$ is taken to be proportional to velocity, as found experimentally for small objects moving slowly through a highly viscous medium.

Open the **Falling Bodies** tool. By adjusting the sliders for m and b, you can change the graph of the velocity $v(t)$. The corresponding motion of the bead is shown in the animation. The bead is assumed to be released at rest: $v(0) = 0$.

1.1   Sketch the graph of $v(t)$ for a few different choices of m and b.

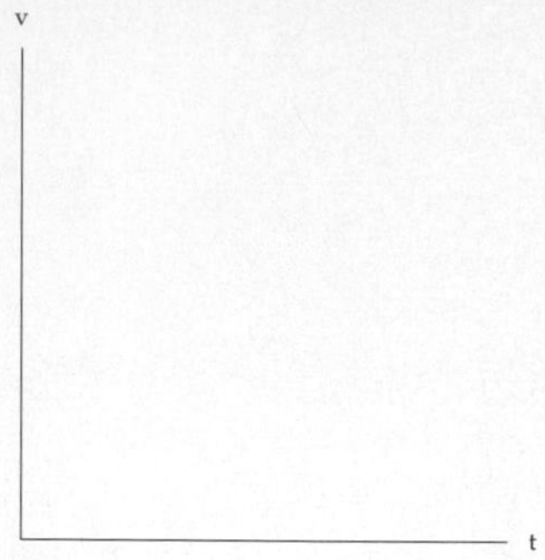

v

t

1.2   The velocity appears to approach a limiting value as $t \to \infty$ (known as the **terminal velocity**). Find a formula for the terminal velocity.

1.3   Describe how the graph of $v(t)$ changes as you increase the mass m.

1.4   Assuming that the bead is released from rest, solve for $v(t)$.

1.5   Find the time required for the bead to reach 50% of the terminal velocity. Does this time increase or decrease as you increase the viscosity b? Give a physical explanation.

1.6   Notice that the graph of the height $y(t)$ starts out curved, and then eventually becomes almost straight. Explain this, and find the slope of the straight part of the graph.

2. How to Hit a Golf Ball as Far as Possible

Herman Erlichson (1983) considered the question, "What angle do you need for maximum projectile range if you're not in a vacuum?" He became interested in this question because, as an avid golfer, he knew that the optimal angle for a golf tee shot is about 11 degrees from the horizontal. This is much less than the optimal launch angle of 45 degrees predicted from calculus, which is based on the simplifying (here, over-simplifying) assumption that air resistance can be neglected.

The first issue is how to model the drag force on a golf ball. There is some controversy here. At the time that Erlichson wrote his article, the available experiments indicated that the drag force on a golf ball is proportional to the velocity, over the range of typical velocities encountered in practice. More recent work (MacDonald and Hanzely, 1991) suggests that the drag force increases with the square of the velocity. As a further refinement, one should also include the aerodynamic lift on the ball, due to its backspin, but we will omit that effect here—students interested in these more realistic cases should look up the papers by Erlichson (1983) and MacDonald and Hanzely (1991).

Assuming that the drag is proportional to velocity, and neglecting lift, Newton's law yields

$$m\frac{dv_x}{dt} = -bv_x$$

$$m\frac{dv_y}{dt} = -bv_y - mg$$

where v_x and v_y are the horizontal and vertical components of the velocity, and the convention is that y and v_y are now measured positive upward. The initial velocity is

$$v_x(0) = s\cos\theta$$

$$v_y(0) = s\sin\theta$$

where $s = 200$ ft/sec is a typical launch speed of a good drive, and θ is the launch angle. The initial position is $x = y = 0$, and the velocity and position are related by $\frac{dx}{dt} = v_x$ and $\frac{dy}{dt} = v_y$. With this information, the flight path of the golf ball is completely determined, given the values of the parameters m, g, b, s, θ.

The values of m, g, b, s can be measured experimentally; thus θ is the only adjustable parameter.

Furthermore, by dividing through by m in the equations of motion, we can see that the separate values of b and m are not important; only their ratio b/m matters.

2.1   Open the **Golf** tool. Using the slider for the initial angle θ, find the angle that maximizes the range of the shot, assuming that $b/m = 0.25 \text{ sec}^{-1}$ and the initial speed is 200 ft/sec. What is the approximate maximum range?

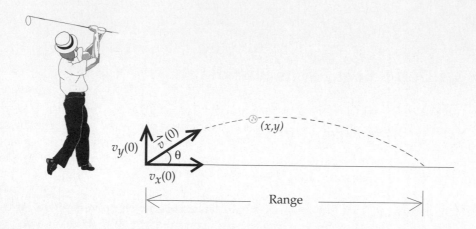

2.2   The experimentally observed optimum angle is about 11 degrees, much less than the answer you should have found for the previous question. What simplifications in the model might be responsible for this discrepancy?

2.3   The predicted flight paths are qualitatively different from those obtained in the absence of air resistance. What is the main difference? At which angles is this effect most pronounced?

2.4   Examine the effects of changing the drag parameter b. If you increase b, how does the optimal angle change? What happens to the maximum range?

2.5   To check that the tool is working correctly, examine the case where $b = 0$. What do you predict for the maximum range? Does this agree with the curves shown by the tool?

2.6   If you wanted to maximize the amount of time that the ball spends in the air, how would you set the sliders for the launch angle, the initial speed, and the air resistance?

2.7   The graph in the lower left panel shows the horizontal velocity x', the vertical velocity y', and the speed $s = \sqrt{(x')^2 + (y')^2}$, all as functions of time. The picture suggests that the graph of $s(t)$ intersects the graph of $x'(t)$ at exactly one point. Give a mathematical argument for why this must be true. In other words, explain why the two curves must intersect once, but cannot intersect more than once.

2.8   Prove that at the point where the graphs of $s(t)$ and $x'(t)$ intersect, the two curves must be tangent to each other.

2.9   Now look at the graph of the height y vs. t. Does it take longer for the ball to go up or down? In other words, which takes more time: the rise to the maximum height, or the fall back to the ground? Explain why the answer makes sense physically.

3. For Further Exploration

There are many other interesting applications of differential equations in connection with the physics of sports. How can a downhill ski racer move faster than a skydiver? What makes a curve-ball curve? Why does a well-thrown football keep its axis pointed along its trajectory? See Armenti (1992) for a fascinating collection of articles that address these and other questions. In many cases, the answers are tentative and controversial; there are opportunities for creative investigations here!

A good class project could involve thinking about the problem faced by the designers of the Prell shampoo commercial. You want the bead to drop so slowly that it takes almost the whole commercial—about 30 seconds—to reach the bottom. What should be the mass of the bead?

To solve this, you'll need to do some experiments with Prell. What measurements do you need to perform to determine the value of the viscosity parameter b? About how tall is a bottle of Prell shampoo? Are there any other measurements you need to make?

For more on the specific subject of projectiles and falling bodies, along with some ideas for class projects, see Minton (1994) and Gruzka (1994).

References

Armenti, Angelo Jr., ed. *The Physics of Sports.* New York: American Institute of Physics, 1992.

Erlichson, Herman. "Maximum Projectile Range with Drag and Lift, with Particular Application to Golf." *American Journal of Physics* 51: 357–362 (1983).

Gruzka, Thomas. "A Balloon Experiment in the Classroom." *College Mathematics Journal* 25: 442–444 (1994).

MacDonald, William M. and S. Hanzely. "The Physics of the Drive in Golf." *American Journal of Physics* 59: 213 (1991), and see the references cited therein.

Minton, Ronald. "A Progression of Projectiles: Examples from Sports." *College Mathematics Journal* 25: 436–442 (1994).

Lab 4:　Tool Instructions

Falling Bodies Tool

Setting Parameters and Initial Conditions

Use the mouse to select the desired parameters and initial conditions by clicking and dragging the two sliders at the upper right (m, b).

Use the mouse to select the initial height of the object by moving the mouse over the bar area, which is located to the left of the first illustration. Click the mouse in the bar area to select the desired height of the object.

Other Buttons

Click the mouse on the [Clear] button to remove all trajectories from the graphing plane.

Golf Tool

Setting Parameters and Initial Conditions

Use the mouse to select the desired parameters and initial conditions by clicking and dragging the three sliders at the lower right (Initial Angle, Initial Speed, and Air Resistance).

Swing Button

Click the mouse on the [Swing] button or the xy plane after setting the initial conditions to see the output.

Other Buttons

Click the mouse on the [Clear] button to remove all trajectories from all graphing planes.

5 Numerical Methods

*How does a computer **solve** a differential equation?*

1. Time Steps

As we've mentioned in previous labs, some differential equations cannot be solved analytically. But all is not lost—we can try to draw a picture of the solution, using the slope field as in Lab 2, or, as shown in this lab, we can use a computer to produce an accurate numerical approximation to the solution, *using nothing more than arithmetic!* (There's no need for calculus or even algebra.) The **Time Steps** tool shows how this works.

To keep things simple, we pick a problem where we (who *do* know calculus!) can easily figure out the exact answer, and see if the computer method is clever enough to find a good approximation to it. For instance, suppose this initial value problem is

$$\frac{dx}{dt} = 2t \text{ , subject to the initial condition } x = 0 \text{ at } t = 0.$$

1.1 Find the exact solution to this problem.

The computer tries to solve this by following the slope field. The idea is to step forward in time by a very small amount Δt, and then ask how much x changes. Since the local slope is $\frac{\Delta x}{\Delta t} \approx \frac{dx}{dt}$, the change in x should be $\Delta x \approx \frac{dx}{dt} \Delta t$, which becomes $\Delta x \approx 2t\Delta t$ for our problem. So the rule is: given some point (t, x), move along the slope field to the new point $(t + \Delta t, x + \Delta x)$, where $\Delta x \approx 2t\,\Delta t$. Then repeat this process, starting from the new point. Keep marching along the slope

37

field in this way to trace out the approximate solution. This is the computer's strategy in **Time Steps**. It is called the *Euler method*,

$$x_{n+1} = x_n + \Delta x_{n'} \text{ or}$$
$$x_{next} = x_{now} + \Delta x_{now}$$

which is the simplest method for *numerically integrating* a differential equation.

1.2   Click on the $\Delta t = 1$ button in **Time Steps**. The computer takes a giant step from $t = 0$ to $t = 1$, following the slope field at $t = 0$. Then at $t = 1$, the computer again computes the local slope, and takes another giant step to $t = 2$.

What is the local slope at $t = 0$? At $t = 1$?

1.3   What is the computer's prediction of x when $t = 1$? How close is that to the true answer that you found above? What about $t = 2$?

1.4   Redo the numerical integration, but now click on the $\Delta t = \frac{1}{2}$ button in **Time Steps**. In what ways is this approximate solution different from the one produced when $\Delta t = 1$?

1.5   Now click on successive smaller step sizes Δt. Why does the approximation improve as Δt decreases?

1.6   If the step Δt is smaller than that shown here, can the numerical approximation ever lie *above* the exact solution, given by the parabolic curve? Explain.

2. Comparing Different Numerical Methods

As you've seen above, the Euler method does a pretty lousy job of approximating the solution unless Δt is extremely small. The **Numerical Methods** tool introduces you to some methods that do better. They are trickier to understand, but much more efficient and accurate.

Open up the **Numerical Methods** tool and click on *Forward Euler*. This uses the Euler method to approximate the solution to

$$\frac{dx}{dt} = 3t^2 \text{, subject to the initial condition } x = 0 \text{ at } t = 0.$$

The method is called *Forward Euler* because it uses steps whose slopes are given at the left hand endpoint of every step of length Δt. In contrast, try *Backward Euler*—this uses the slope at the right endpoint of each step.

2.1 Suppose we had used $\Delta t = 0.25$ instead of $\Delta t = 0.5$. What value would *Forward Euler* predict for x when $t = 0.5$? Remember that $\Delta x = \left(\dfrac{dx}{dt}\right)\Delta t$.

Symbolically rewrite Δx so that it applies to this example only:

Now use the following tables to answer the question about what *Forward Euler* will predict, and do the same for *Backward Euler*.

Forward Euler

t	x	Δx	new x

Backward Euler

t	x	Δx	new x

2.2 In this example, we see that *Forward Euler* gives an approximation that lies below the true curve, whereas *Backward Euler* lies above. Sketch the graph of a function $x = f(t)$ where the opposite is true.

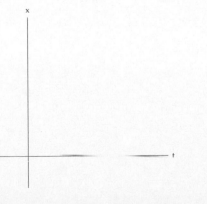

2.3   The graphs suggest that if we could somehow average *Forward Euler* and *Backward Euler,* we'd get a better approximation. The *Trapezoidal* (*improved Euler*) method takes the slopes predicted by the *Forward Euler* and *Backward Euler* methods, and then averages those slopes. Click on the *Trapezoidal* button to see what that approximation looks like, and sketch the result.

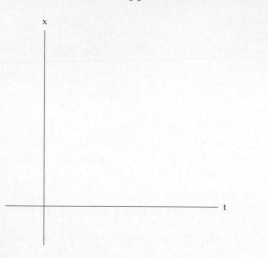

2.4   The *Midpoint* method tries a different strategy: it uses the slope at the midpoint of a step, rather than at the left or right endpoint. For the example we are considering, which method produces a generally better approximation, *Midpoint* or *Trapezoidal?*

2.5   For the best method so far, we take an unequal mixture of *Midpoint* and *Trapezoidal.* The popular *Runge-Kutta* method (in the simplest case, sometimes called *Simpson's*) is defined as

$$\frac{2}{3}(Midpoint) + \frac{1}{3}(Trapezoidal).$$

Why is it plausible that this should be better than simply averaging *Midpoint* and *Trapezoidal* equally?

3. How the Error Depends on the Stepsize

Our earlier work with the **Time Steps** tool showed that the Euler method gives an increasingly accurate approximation to the true solution as we decrease the stepsize Δt. The same trend holds for all the other methods introduced in the **Numerical Methods** tool. That makes sense, since smaller steps always mean a better approximation of the derivative, no matter what method we use. The interesting twist is that some of the methods provide much more "bang for the buck" as the stepsize is decreased. For instance, we'll see below that a tenfold decrease in Δt yields about a tenfold increase in accuracy for the Euler method, but a ten*thousand*fold increase in accuracy for the Runge-Kutta method! Our goal now is to investigate this dramatic difference.

Open the **Numerical Methods: Stepsize Scaling** tool. To change the stepsize, click on one of the data points in the graph on the right, and you'll see that the curves are redrawn in the graphs on the left. Let's now explain what these graphs are showing.

The top graph on the left shows three curves. The gray curve is the exact solution $x(t) = t^5$ of the equation $\dfrac{dx}{dt} = 5t^4$, subject to the initial condition $x(0) = 0$. The yellow curve is the approximate solution given by the (forward) Euler method. (When you first open the tool, the default stepsize is $\Delta t = 1$.) The blue curve is the approximate solution given by the Runge-Kutta method, also with the same stepsize.

Note that for $\Delta t = 1$ the Runge-Kutta points are almost indistinguishable from the exact solution, whereas the Euler method produces a visible discrepancy at each step.

To compare the methods quantitatively, let's look at the error, which is defined as the absolute value of the difference between the true solution and the numerical solution at each value of t. The error is plotted as a function of t in the bottom graph on the left. The red and green numbers are the errors at $t = 3$ for the Euler and Runge-Kutta methods, respectively.

3.1   Estimate the value of the error for the Euler method at $t = 1$, $t = 2$, and $t = 3$, using a stepsize $\Delta t = 1$.

3.2   Explain intuitively why the error increases with time.

Although it is interesting to observe how the error grows with time, we are more concerned with the error as a function of the stepsize Δt. The graph on the right shows a plot of $\log_{10}$ (error) measured at the right-hand endpoint $t = 3$, plotted as a function of $\log_{10}(\Delta t)$. The reason for using logarithms on both axes (known as a log-log plot) is that the data then fall on a straight line, as you can see. The five different data points correspond to different stepsizes, and the red and green colors correspond to the Euler and Runge-Kutta methods, respectively.

To decrease the stepsize, click on one of the data points farther to the left.

3.3   What are the five stepsizes used in this plot?

3.4   What is the approximate slope of the line corresponding to the Euler data?

3.5   What is the approximate slope of the line corresponding to the Runge-Kutta data?

3.6   Let E denote the error measured at $t = 3$. Assume that E is related to Δt by a relation of the form $E \approx C(\Delta t)^n$, where C is some positive constant and n is a positive integer. Show that if $\log_{10} E$ is plotted vs. $\log_{10}(\Delta t)$, the graph is a straight line with a slope equal to n.

3.7   The slope n is called the **order** of the numerical method. What are the orders of the Euler and Runge-Kutta methods? (You can assume that the data given here are representative of the methods' performance on other examples.)

3.8   Show that if we decrease Δt by a factor of 10, the resulting error E decreases by a factor of about 10^n.

Lab 5: Tool Instructions

Time Steps Tool

Buttons

Click on a [Δt] button below the graphing plane to set a value for the time step.

Numerical Methods Tool

Buttons

Click on one of the five buttons below the graphing plane to choose a numerical method.
Click the mouse on the [**Clear**] button to remove all numerical solutions from the graphs.

Numerical Methods: Stepsize Scaling Tool

Step Conditions

Click the mouse in the Stepsize window to choose a stepsize interval from 0.01 to 1.00.

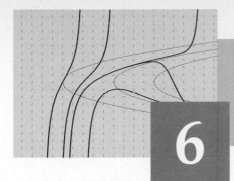

6 Isoclines and Fences

Tools Used in Lab 6

Isoclines

Isoclines as Fences

To the Instructor: Why bother with this topic? Traditional courses often omit isoclines, and only the newest texts might refer to fences and funnels. But both concepts are important in moving toward a useful geometric view of differential equations. Now that graphics have helped us to make the pictures part of a student's "world view" of differential equations, isoclines can be valuable in making a quick sketch when one is not at a computer or calculator. Fences and funnels make a simple and consistent way to approach many a proof, including many that actually attract students. See Section 2 for a bit of illustration.

Solutions in a slope field wander from isocline to isocline, leaping over fences, sometimes slipping out of antifunnels, sometimes sliding into funnels, never to escape! What are these notions?

1. Isoclines

An **isocline** is a curve on which the solutions to the differential equation have the same slope. For a given differential equation $\dot{x} = f(t, x)$, set the function $f(t,x)$ equal to a constant to define an isocline.

$$\dot{x} = f(t,x) = c$$

For any particular value of the constant c we usually obtain an equation for a curve, the isocline. However, there may be values of c for which no such curve exists.

1.1 Give an example of $\dot{x} = f(t, x)$ where $f(t,x) = c$ does *not* exist for some c. If $c = 0$, for example, can you find a function $f(t,x)$, where $f(t,x) = 0$ does not exist?

Why Are Isoclines Important?

Before the widespread availability of computer solvers for differential equations, isoclines were used as an aid to graphing solutions. They are often the easiest way to sketch a solution by hand. Now they are more important as a check on the reliability of computer-generated solutions and as a way to gain insight into the long-term behavior of solutions (see Section 2). Remember that isoclines are *not* solutions (except in some unusual instances to be discussed later), so it is always a good idea to draw them in a different color than the solutions.

1.2 Using the **Isoclines** tool, choose the differential equation $\dot{x} = x^2 - t$, and, using the "slider" for c, find some isoclines. Include isoclines for $c = 0$ and for several positive and negative values of c. Note that the slope marks on each isocline denote the direction of the slope vectors of the solutions passing through the isoclines.

From your knowledge of the fact that solutions move tangent to the slope marks, try to predict the path of the solution through the point (1,0). Now click on that point to find the solution through (1,0). How did you do?

1.3   Try out the different equations in the **Isoclines** tool to see different isocline and solution behaviors. For each of the equations below, sketch a few of the solutions. Then write a short description of what the isoclines look like, and how the solutions behave.

$$\dot{x} = x(1-x) \qquad\qquad \dot{x} = t - 2x \qquad\qquad \dot{x} = x^2 - t^2$$

A nullcline is an isocline for which $\dot{x} = 0$, that is, $f(t,x) = 0$. A nullcline sketch makes the sketching of solution curves much easier, and can be used to prove the behaviors of solutions. With a colored pen or pencil, locate the nullclines on the graphs for each of the equations in the **Isoclines** tool.

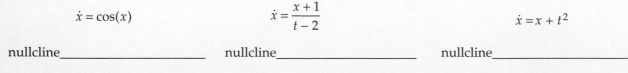

1.4   Find the nullcline formula for each differential equation below. Use colored pencils to sketch the nullcline and at least three isoclines on each slope field.

$$\dot{x} = \cos(x) \qquad\qquad \dot{x} = \frac{x+1}{t-2} \qquad\qquad \dot{x} = x + t^2$$

nullcline_____   nullcline_____   nullcline_____

1.5   Can you find an exception to the general rule that "an isocline is not a solution"? Look at the examples from Exercise **1.3**. Confirm your finding algebraically.

2. Fences, Funnels, and Antifunnels

The ideas in the remainder of this lab—fences, funnels, and antifunnels—were developed by John Hubbard at Cornell University to make simple but rigorous proofs about the long-term behaviors of solutions to differential equations. They provide a vocabulary that allows easy descriptive statements and short proofs rather than lengthy arguments. Such qualitative techniques often provide surprisingly quantitative information, such as location of asymptotes to any desired accuracy. A detailed discussion can be found in [HW], Sections 1.3 through 1.6 and 4.7, with examples and applications beyond those discussed below.

Fences

Turn to the **Isoclines as Fences** tool, change the constant with the slider, and notice the different colors used to draw the isocline. The two colors denote **upper fences** and **lower fences**. A **fence** is some curve other than a solution that "channels" the solutions in the direction indicated by the slope field. Isoclines can be fences, but so can many other curves, as we discuss in the exercises.

Look at the screen for $\dot{x} = x^2 - t$. If we consider points of intersection between solutions and another curve (in this instance an isocline), the curve is an upper fence if the solution slopes are *less* than the curve slopes at the intersections, and is a lower fence if the solution slopes are *greater* than the curve slopes. Solutions move up from lower fences and down from upper fences, as they go from left to right.

2.1   Note that isoclines are not the only possible fences. Consider the slope field for $\dot{x} = 2x^2 - 1$ with some solutions sketched. With a pencil, draw a horizontal line and a couple of oblique lines across the picture. Then color parts of the lines according to whether they act as upper or lower fences.

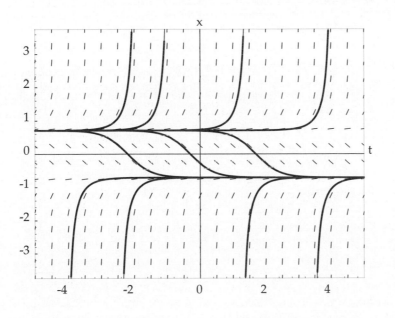

Funnels and Antifunnels

Visualize solutions converging toward a single solution as $t \rightarrow \infty$. We say the solutions are **funneled** toward the single solution and we describe the pattern of solution behaviors as a **funnel.** If an upper fence is always above a lower fence, the region inside is a funnel.

Alternatively, imagine solutions diverging or moving away from an exceptional solution. We describe this pattern of solution behaviors as an **antifunnel.** If an upper fence is always below a lower fence, the region inside is an antifunnel.

2.2   The fences for some funnels and antifunnels are sketched below. Which pairs are which? Label the sketches appropriately.

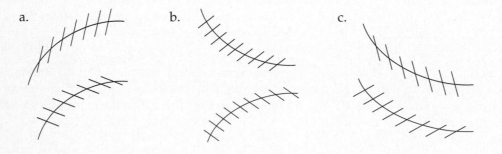

a.   b.   c.

2.3   Some solution curves for $\dot{x} = x^2 - t$ are shown below, with isoclines shown as finer lines. With colored pencils, shade and label a narrowing funnel and a narrowing antifunnel.

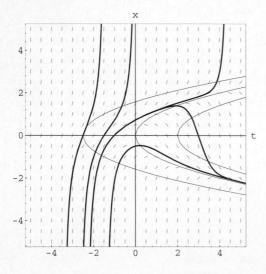

A unique solution gets trapped in the narrowing antifunnel you have just highlighted. Add to your picture a sketch of the unique solution that separates solutions that go to positive infinity from those that converge in the funnel.

Defining the locations of the funnel and antifunnel allow us to describe the long term-behavior of solutions more precisely than "some of them blow up" or "others approach negative infinity"—we know exactly *how* they do this. We can say more clearly that "for $x > \sqrt{t}$, solutions blow up (apparently with vertical asymptotes), and for $x < \sqrt{t}$, solutions are funneled toward negative infinity at the rate of $-\sqrt{t}$."

2.4   Make an analysis similar to Exercise **2.3** for at least one of the other equations in the **Isoclines as Fences** tool. Give a rough sketch and describe your results in the space below.

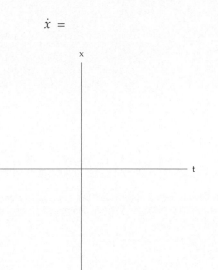

$$\dot{x} =$$

Why else, besides allowing precise descriptions, might fences and funnels be important or useful? Another example comes from the fact that pictures of approximate solutions may be misleadingly incorrect! For instance, the t-domain might be too large for the stepsize used in calculation, or the stepsize might be too large for the domain under study. Then you would want to recognize a mistake, and/or confirm a correct picture.

2.5   The equation $x' = 1 - tx$ is awkward at best to solve analytically, but it is easy to make pictures of the families of solutions. Each of the following pictures, however, is *wrong*, due to too big a stepsize (0.5) over too big a domain (with t going to 7).

Euler's Method   Runge-Kutta Method

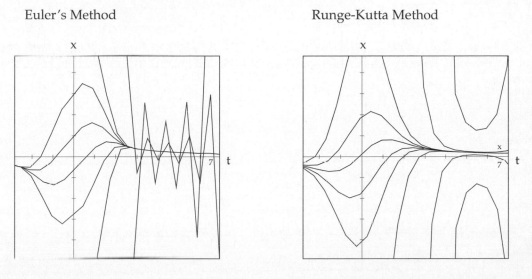

    a.   You should have an immediate clue that the picture from Euler's method on the left has at least some solutions that are not correct. Give a reason why.

b. Label each of the Euler "solutions" with R for those that are approximately right and W for those that go wrong. Use the isoclines of slopes 0 and 1 to prove your answers, and indicate *where* the W solutions go wrong.

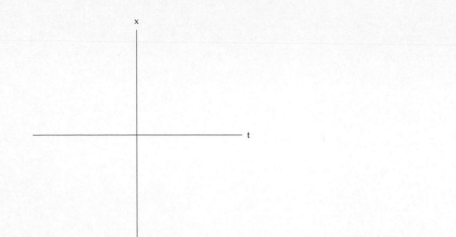

c. From a first glance at the right-hand picture from the Runge-Kutta method, it is not obvious that there might be anything wrong, unless you have been thinking about the isoclines, or some other helpful aspect! If so, tell us, but don't worry if you have nothing to say here.

d. Label each of the Runge-Kutta "solutions" with R for those that are approximately right and W for those that go wrong. Add to your graph the isoclines of slopes 0, 1, and −1. (You'll want to think about other isoclines as well, but these provide the core of the argument.) Then make a fence and funnel argument to prove your answers, and indicate *where* the W solutions go wrong.

e. Make a sketch of the correct solutions, with the isoclines of slopes 0, 1, and −1. If you can, make observations or remarks due to additional information, e.g., from efforts to solve the equation.

Moral: You cannot just believe a picture of a numerical approximation. In IDE we have tried to control the stepsizes and domains to avoid this kind of mistake. But when you are on your own using an open-ended graphing tool, with any old equation, you cannot be so sure. Isoclines, fences, and funnels are handy tools for checking out your graphs.

Check: Go back to the **Solutions** tool for Lab 2, Graphing Differential Equations. Note that for Euler with stepsize 0.5 you can get a recognizable error, for certain initial conditions!

3. Additional Exercise

3.1   For the differential equation with the slope field and solutions shown below, add fences that enable you to make an analysis and description of the solutions. These fences can be composed of lines or isoclines, or portions of either. *(Hint:* Find the oblique isocline of horizontal slope; then look to the right of that and try horizontal fences arbitrarily close to some interesting graphical features.)

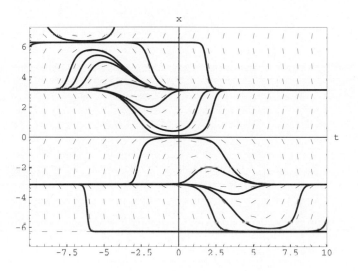

Lab 6: Tool Instructions

Isoclines Tool

Setting Initial Conditions

Click the mouse on the graphing plane to set the initial condition for a trajectory.

Clicking while a trajectory is being drawn will stop the trajectory and draw a new one.

Parameter Sliders

Use the slider to set the constant slope c and reposition the isocline.

Press the mouse down on the slider knob and drag the mouse left and right, or click the mouse in the slider channel at the desired value for the parameter.

Equations

Click the button to the left of the equation to scroll the list of equations.

Click an equation to select it.

Buttons

Click the mouse on the [Clear] button to remove all trajectories from a graph.

Click the mouse on the [Draw Field] button to draw a slope field over the graphing plane.

Isoclines as Fences Tool

Setting Initial Conditions

Click the mouse on the graphing plane to set the initial condition for a trajectory.

Clicking while a trajectory is being drawn will stop the trajectory and draw a new one.

Parameter Sliders

Use the slider to set the constant slope c and draw a new isocline.

Press the mouse down on the slider knob and drag the mouse left and right, or click the mouse in the slider channel at the desired value for the parameter.

Equations

Click the button to the left of the equation to scroll the list of equations.

Click an equation to select it.

Buttons

Click the mouse on the [Clear] button to remove all trajectories from a graph.

Click the mouse on the [Draw Field] button to draw a slope field over the graphing plane.

Existence and Uniqueness

7

Tools Used in Lab 7

Targets
Sure-Fire Target
Uniqueness

How do we know that a solution to a differential equation exists for a given initial value? Myriad solutions may exist, sometimes radiating star-like from a particular point, or there may be only one solution. In some cases there are none.

Target Shooting

Target shooting leads in a natural manner to concepts of existence and uniqueness of solutions to first order differential equations. There is, however, a limitation to the **Targets** tool that arises from the fact that a pixel on the screen has a discrete non-zero area. In the neighborhood of converging solutions a target may be hit by more than one trajectory. This spurious effect has been disguised somewhat by the tool designer.

Open the **Targets** tool. Select the differential equation $\dot{x} = x^2 - t$. Pick a point in the field to be a target and click on it with the mouse. Pick another point some distance away that you think may lie on the same solution curve. "Shoot" at the target by clicking on the second point. Keep selecting points from a sporting distance until you get a hit. The distance that is "sporting" varies with the behavior of solution curves in the neighborhood of the target. If the solution curves are diverging you may need to move quite close to the target to get a hit. If the solution curves are converging toward the target, you will be able to hit the target from much further out. To select a new target point, click the **Clear Target** button, then click the mouse on the desired spot. Try some other equations, especially the "tricky ones."

Two Important Questions (Not to answer now, but to keep in mind as you are doing this lab.)

1. **Can you always hit the target?** Is there always at least one solution through a given target point? That is, does a solution exist?

2. **Can you hit the target more than once?** Can there be more than one solution through a given target point? That is, is a solution unique?

1. Existence

The first important question is the question of **existence**.

Given $\dot{x} = f(t, x)$ and a point on the plane, is there a solution that passes through the point? The answer is *yes* if $f(t, x)$ is real-valued and *continuous in an open region containing the point*.

1.1   Find at least two examples from the differential equations in the **Targets** tool where there are large regions in which no solutions exist. Describe the regions for each equation.

1.2   Consider the differential equation $\dot{x} = x / \cos(t)$. For what values of (t, x) can solutions *not* be guaranteed?

2. Uniqueness

The second important question is the question of **uniqueness.**

Given $\dot{x} = f(t, x)$ and a point on the plane, is a solution that passes through the point the only such solution? Now we add a new "smoothness" condition on $f(t, x)$ such as *continuous differentiability throughout an open region containing the point*. To experience non-uniqueness, try the **Sure-Fire Target** tool.

2.1   Start with the differential equation $t\dot{x} = x$. From your observations, for what point or points in the plane do solutions exist? For what point or points are the solutions unique? (*Hint*: Sometimes you may be shooting "backwards in time"!)

2.2   Solve the differential equation $t\dot{x} = x$ analytically.

Did you find a solution that passes through $(t, x) = (1, 0)$? What would it be? Note that $x \equiv 0$ is a valid solution. Modify your answer to Exercise **2.1** if necessary.

2.3   Try $t\dot{x} = x + t^2 \cos(t)$ in the **Sure-Fire Target** tool. What kind of differential equation is this? Is it separable, exact, or linear? Solve this equation. Obtain all the solutions.

2.4   For what points in the plane are there unique solutions? non-unique solutions? no solutions? Show, with your "family of solutions" in Exercise **2.3**, that any non-unique solutions you might have are justified.

2.5   Open up the **Uniqueness** tool. Consider the two examples $\dot{x} = x^{2/3}$ and $\dot{x} = x^{4/3}$. For what regions (for example, for $a < t < b$, $c < x < d$) can the existence of solutions be guaranteed for an initial point in the region? How must we further restrict the regions to be certain that the solutions are unique? For the analysis, check to see where, if at all, the functions $f(t, x)$ fail to be continuously differentiable (which requires that $\partial f / \partial x$ exists and is continuous).

The pictures for both differential equations are a bit tricky. To see the behavior of the solutions more clearly as they approach the x-axis, look at the vertical enlargement shown on the right. Note that one differential equation has distinct solutions everywhere, some of which are approaching the horizontal axis asymptotically, and the other has non-unique solutions on the horizontal axis. These results should agree with your analysis.

2.6   The computer is somewhat selective about which solution it shows when there are many solutions through a point, say (1,0). If you pick a point on the horizontal axis, the solution follows the horizontal axis through (1,0); there are, however, many other solutions through this point. Note that solutions may stay unique in one region but be non-unique in larger regions. Remember, the solution is any smooth (differentiable) function that passes through (1,0) and satisfies the equation.

What Happens as a Solution Crosses the t-Axis?

If you pick a point that is off the horizontal axis, the computer only shows the solution that crosses directly over the axis in the direction it was heading. However, there may be infinitely many ways for a solution to continue when it reaches the axis. For example, the solutions to $\dot{x} = x^{2/3}$ through $(1,0)$ may be obtained by hooking up pieces of solutions joined together (in proper fashion) by a horizontal piece along the t-axis containing $(1,0)$. Using colored pencils, sketch three solutions through $(1,0)$.

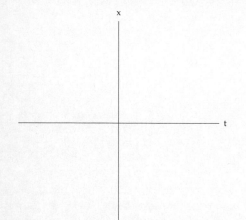

2.7 Solve both of the following equations analytically. Compare the solutions with the graphical results. Is $x \equiv 0$ a valid solution for both? For each equation, are there other solutions that intersect the t-axis? What does this say about the uniqueness of $x \equiv 0$?

$$\dot{x} = x^{2/3}$$

$$\dot{x} = x^{4/3}$$

2.8 Determine $\dfrac{\partial f}{\partial x}$ for both examples in Exercise **2.5**. Are your results consistent with the uniqueness test at the beginning of Section 2?

The questions of existence and uniqueness are not trivial mathematical questions. In fact, it is important to realize that most differential equations do *not* have explicit formulas for their solutions. Nevertheless, you have learned that with graphics, if solutions exist, you can see them, describe them, and make predictions about their behavior in the long term.

3. Additional Exercise

Note that even if a unique solution exists at every point, you might have a case where the solutions "blow up"—become infinite in a finite time. Look at the example $\dot{x} = 1 + x^2$ using the **Targets** tool. Solve it by separation of variables to obtain the analytical solution. Use this to explain what happens if you try to "shoot" at $(0, 0)$ from (a, b) where $a > \dfrac{\pi}{2}$.

Lab 7: Tool Instructions

Targets Tool

Setting Initial Conditions

The first click on the graphing plane will set a target. Once a target is set, click on another point to set the initial conditions $x(0)$ and $t(0)$ for a solution curve that you think will pass through the target point. Clicking while a trajectory is being drawn will start a new trajectory.

Equations

Click the button to the left of the equation to scroll the list of equations.
Click an equation to select it.

Buttons

Click the mouse on the **[Clear]** button to remove all trajectories from a graph.
Click the mouse on the **[Clear Target]** tool to remove the active target point, then click on the plane to set a new target.
Click the mouse on the **[Draw Field]** button to draw a grid of vectors over the graphing plane.

Sure-fire Target Tool

Setting Initial Conditions

Click the mouse on the graphing plane to set the initial condition for a trajectory.
Clicking while a trajectory is being drawn will start a new trajectory.

Equations

Click the button to the left of the equation to scroll the list of equations.
Click an equation to select it.

Buttons

Click the mouse on the **[Clear]** button to remove all trajectories from the graph.
Click the mouse on the **[Draw Field]** button to draw a slope field over the graphing plane.

Uniqueness Tool

Setting Initial Conditions

Click the mouse on the graphing plane to set the initial condition for a trajectory.
Clicking while a trajectory is being drawn will start a new trajectory.

Equations

Click the button to the left of the equation to scroll the list of equations.
Click an equation to select it.

Buttons

Click the mouse on the **[Clear]** button to remove all trajectories from the graph.
Click the mouse on the **[Draw Field]** button to draw a slope field over the graphing plane.

8 Orthogonal Trajectories

Tools Used in Lab 8

Orthogonal Trajectories

If two families of curves always intersect each other at right angles, then they are said to be orthogonal trajectories of each other. Given one family, can you predict the other?

1. Orthogonal Curves

Try a few equations in the **Orthogonal Trajectories** tool. Note that as a point on the plane is selected and clicked on, both the curve (yellow) and its orthogonal trajectory (blue) are given through the point.

If one family consists of the solution curves for

$$\frac{dy}{dx} = f(x,y),$$

then the family of orthogonal trajectories must be solution curves with orthogonal slopes. This means that they must satisfy a different differential equation:

$$\frac{dy}{dx} = \frac{-1}{f(x,y)}.$$

This is the general method. If we start with the equation for the family of curves, we must first differentiate to find the differential equation for the slopes, then find the differential equation for the orthogonal slopes.

1.1 Consider the family of curves $xy = C$. The curves are pictured below. On the same graph, sketch the orthogonal family. After you have drawn a few of these curves you will find it simple to predict orthogonal trajectories.

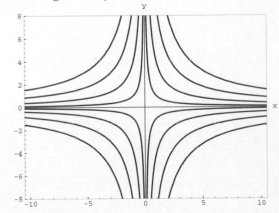

63

1.2 Find the equations for the orthogonal family of curves using analytical methods.

Hint: You should first use implicit differentiation on $xy = C$ to find $\dfrac{dy}{dx} = f(x, y)$ for the original curves. Then solve a new differential equation, $\dfrac{dy}{dx} = -\dfrac{1}{f(x, y)}$.

Now check your sketch with the **Orthogonal Trajectories** tool. Look at the list of differential equations to find the one with solution curves $xy = C$. Does your sketch agree? If not, you can fix it now.

1.3 As in Exercise **1.1**, visualize and sketch the graph of the orthogonal trajectories of the logistic equation: $\dfrac{dy}{dx} = y(1-y)$. Then check your prediction with the **Orthogonal Trajectories** tool.

What is the differential equation that the orthogonal trajectories must satisfy?

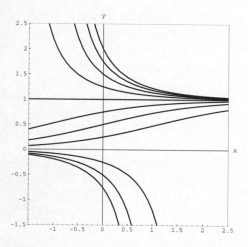

The **Orthogonal Trajectories** tool generates *two* families of mutually orthogonal curves, yellow and blue, respectively; each family can be found if you know the other.

In fact, most ODE solvers allow us to graph orthogonal trajectories in the following way. Start with the differential equation $\dfrac{dy}{dx} = \dfrac{h(x, y)}{g(x, y)}$. Define $\dfrac{dx}{dt} = g(x, y)$, $\dfrac{dy}{dt} = h(x, y)$ where x and y are viewed as functions of t. Now the trajectories in the *phase plane*, the (x, y)-plane, are the solution curves for the original differential equation $\dfrac{dy}{dx} = \dfrac{dy/dt}{dx/dt} = \dfrac{h(x, y)}{g(x, y)}$. We can redefine $\dfrac{dx}{dt} = -h(x, y)$ and $\dfrac{dy}{dt} = g(x, y)$ to obtain the slope $\dfrac{dy}{dx} = \dfrac{dy/dt}{dx/dt} = \dfrac{g(x, y)}{h(x, y)}$ of the orthogonal trajectories which can now be graphed in the phase plane. It is possible to find two mutually orthogonal families of solution curves even though we may not be able to solve either of the corresponding differential equations algebraically. Several such cases appear in the **Orthogonal Trajectories** tool.

1.4   Match the pairs of families of orthogonal trajectories. Note that these are not screen pictures, but are on a larger scale, $-10 \le t \le 10$, $-8 \le t \le 8$, giving a larger view of solutions.

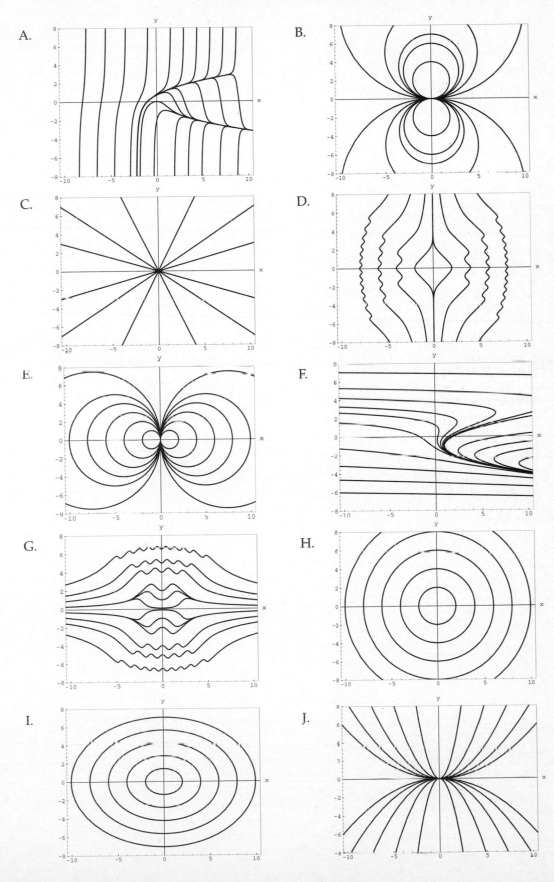

A.

B.

C.

D.

E.

F.

G.

H.

I.

J.

1.5   Match the differential equations with the solution graphs in Exercise **1.4**.

_____ a. $\dfrac{dy}{dx} = \dfrac{2xy}{x^2 - y^2}$ _____ f. $\dfrac{dy}{dx} = \dfrac{y^2 - x^2}{2xy}$

_____ b. $\dfrac{dy}{dx} = \dfrac{-x}{2y}$ _____ g. $\dfrac{dy}{dx} = \dfrac{2y}{x}$

_____ c. $\dfrac{dy}{dx} = \dfrac{-x}{y}$ _____ h. $\dfrac{dy}{dx} = \dfrac{y}{x}$

_____ d. $\dfrac{dy}{dx} = \sin(xy)$ _____ i. $\dfrac{dy}{dx} = -\csc(xy)$

_____ e. $\dfrac{dy}{dx} = y^2 - x$ _____ j. $\dfrac{dy}{dx} = \dfrac{1}{x - y^2}$

2. Some Practical Considerations

The electric field and equipotential lines around a charge are mutually orthogonal families of curves in the plane.

2.1   Which one of the families of curves in Exercise **1.4** represents the electric field lines around a point charge represented by a point in the plane? In three dimensions, these would be the field lines around a charged wire passing through the origin perpendicular to the plane.

2.2   Which family represents the equipotential curves? As above, these would actually be equipotential surfaces in three-dimensional space.

2.3   Can you identify the pair that represents the magnetic field lines about a dipole and the magnetic equipotentials?

Lab 8: Tool Instructions

Orthogonal Trajectories Tool

Setting Initial Conditions
Click the mouse on the graphing plane to set the initial conditions for a trajectory and its associated orthogonal trajectory.
Clicking while a trajectory is being drawn will start a new trajectory.

Equations
Click the button to the left of the equation to scroll the list of equations.
Click on an equation to select it.

Buttons
Click the mouse on the [Clear] button to remove all trajectories from the graph.
Click the mouse on the [Draw Field] button to draw a slope field over the graphing plane.

II

Second Order Differential Equations

Linear Oscillators: Free Response

9

A mass bobbing on a spring, the quiet ticking of a grandfather clock, a child gently swinging under a tree—how can all these be modeled as unforced linear oscillators by means of homogeneous second-order linear differential equations with constant coefficients?

1. The Undamped Mass-Spring System

Open the **Simple Harmonic Oscillator** tool. The energy in a simple harmonic oscillator is completely determined by the initial position and initial velocity. There is no friction or external forcing. Set a variety of nontrivial initial conditions by clicking on the phase plane, then observe the motion in the phase plane and time series graphs.

The second-order differential equation that describes the mass spring system is

$$m\ddot{x} + kx = 0, \tag{1}$$

where m, k, $\ddot{x}$, and x denote the mass, the spring constant, the acceleration, and the displacement of the mass from its equilibrium position, respectively. The sign of x is positive when the spring is stretched and negative when the spring is compressed.

1.1   In order to graph a second-order differential equation, it is necessary to rewrite it as a system of two first-order equations for velocity, $\dot{x} = v$, and acceleration, $\dot{v}$. Rewrite Equation (1) as

$$\dot{x} = v$$

$$\dot{v} =$$

1.2   As you observe the time series, note the relationship between the acceleration $\ddot{x}$ and the displacement x. Where is acceleration $\ddot{x}$ illustrated? How do you know this? Does this relationship hold for all initial conditions? Explain.

1.3   Describe how the displacement x and the velocity $v = \dot{x}$ are related. For what displacement is the velocity a maximum? a minimum?

1.4   Observe the phase plane. Note that the trajectories on the phase plane are always closed. What does this indicate about the motion?

1.5   Solve Equation (1) for $x(t)$, the displacement as a function of time. What is the frequency, ω, of the oscillation in radians per second?

1.6   Assume $m = 1$ and $k = 1$ for the tool. Calculate ω for these values.

1.7   Energy

a.   With no external forces, the only energy in the system is supplied by the initial displacement and the initial velocity. Look at the energy graph on the **Simple Harmonic Oscillator** tool.

b.   In general, the potential energy is the energy stored by the extension or compression of the spring. For a given displacement x, it is given by

$$E_{potential} = \tfrac{1}{2}kx^2.$$

What is the total energy of the system?

$$E_{total} = E_{potential} + E_{kinetic} = \underline{\hspace{3cm}}$$

c.   Will the total energy in this system dissipate as time increases? Explain.

d.   Calculate the total energy under the following initial conditions, for arbitrary k and m:

$$x(0) = 2$$
$$\dot{x}(0) = -1.$$

2. The Damped Mass Spring System

Open the **Mass and Spring** tool. Set some nontrivial initial conditions and observe the phase plane and the time series. The differential equation now includes a term for the frictional force:

$$F_{friction} = bv,$$

which is assumed to be proportional to the velocity of the mass. This would be the case for viscous damping (frequently illustrated with a "dashpot," a plunger in a cylinder filled with viscous "goo"). The damping constant b is the constant of proportionality. The new equation is linear with constant coefficients:

$$m\ddot{x} + bv + kx = 0 \qquad\qquad\qquad (2)$$

2.1   Rewrite Equation (4) as a system of first-order linear differential equations.

$$\dot{x} =$$

$$\dot{v} =$$

2.2   Describe the effect of damping on phase plane trajectories and the time series. Point out the differences in the graphs from those without damping.

2.3   From Equation (2) we can obtain the characteristic equation:

$$m\lambda^2 + b\lambda + k = 0$$

a.   Show that there are three possibilities: two real unequal negative values for λ, a repeated real negative root for λ, or a complex conjugate pair of roots. Show algebraically that the repeated root occurs when $b = \sqrt{4mk}$. This value of b is denoted b_c, the critical damping.

b. Use the slider to try values of b that are above, at, and below b_c, and observe the graphs. Label the three cases below with the appropriate terms: underdamped, overdamped, and critically damped. Make rough sketches of the time series for $x(t)$ for each of these cases, using the same initial conditions for comparison. State your initial conditions and show $x(0)$ appropriately on the sketch.

$x(0) =$ _____

$v(0) =$ _____

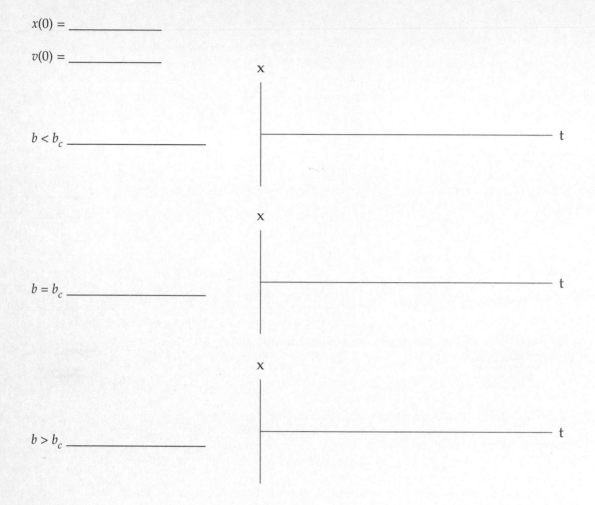

$b < b_c$ _____

$b = b_c$ _____

$b > b_c$ _____

2.4   The Considerate Screen Door. Suppose you wanted to design a damper on a screen door so that it would close as quickly as possible after being released. Due to the intense studying underway by your roommates, you also want it to close gently and smoothly without banging. What is the most desirable amount of damping?

2.5   Now that damping has been included in the model, we have dissipative heat losses due to friction. Open the **Damped Vibrations: Energy** tool and look at the energy graph. Vary the damping constant b to see how the energy loss due to friction is affected. The total initial energy is determined by the initial velocity, $\dot{x}_0 = v_0$, and the initial displacement, x_0,

$$E_{initial} = \tfrac{1}{2}k(x_0)^2 + \tfrac{1}{2}m(v_0)^2$$

How do you determine the energy loss due to friction?

Lab 9: Tool Instructions

Simple Harmonic Oscillator Tool

Setting Initial Conditions

Click the mouse on the $x\dot{x}$ phase plane to set the initial position and the initial velocity.
Click the mouse on the time series graph to set the initial position (initial defaults to zero). Clicking while a trajectory is being drawn will start a new trajectory.

Time Series Buttons

The buttons labeled
  [ ] **position**
  [ ] **velocity**
  [ ] **acceleration**
toggle the time series graphs on and off.

Other Buttons

Click the **[Clear]** button to remove all trajectories from the graphs.
Click the **[Pause]** button to stop a trajectory without canceling it.
Click the **[Continue]** button to resume the motion of a paused trajectory.
Click the **[Energy Graph]** button to automatically display potential energy E_p and kinetic energy E_k whenever a trajectory is drawing the $x\dot{x}$ plane

Mass and Spring Tool

Setting Initial Conditions

Click the mouse on the $x\dot{x}$ phase plane to set the initial position and the initial velocity.
Click the mouse on the time series graph to set the initial position (initial defaults to zero). Clicking while a trajectory is being drawn will start a new trajectory.

Parameter Slider

Use the slider to set the damping constant b, m, and k. The dashpot will disappear when b is set to zero.
Press the mouse down for the parameter you want to change, and drag the mouse back and forth to change it, or click the mouse in the slider channel at the desired value for the parameter.

Time Series Buttons

The buttons labeled
  [ ] **position**
  [ ] **velocity**
  [ ] **acceleration**
toggle the time series graphs on and off.

Other Buttons

Click the **[Clear]** button to remove all trajectories from the graphs.
Click the **[Pause]** button to stop a trajectory without canceling it.
Click the **[Continue]** button to resume the motion of a paused trajectory.
Click the **[Draw Field]** button to draw a slope field over the $x\dot{x}$ graphing plane.

Damped Vibrations: Energy Tool

Setting Initial Conditions

Click the [Start] button to start a trajectory using preset initial conditions.
Clicking in the time series will set an initial value of x and start a new trajectory.

Parameter Slider

Use the slider to set the damping coefficient b.
Press the mouse down for the parameter you want to change, and drag the mouse back and forth to change it, or click the mouse in the slider channel at the desired value for the parameter.

Time Series Buttons

The buttons labeled
 [ ] **position**
 [ ] **velocity**
 [ ] **acceleration**
toggle the time series graphs on and off.

Other Buttons

Click the [Pause] button to stop a trajectory without canceling it.
Click the [Continue] button to resume the motion of a paused trajectory.

10 Free Vibrations

A mechanic pushes down on the front end of a car to observe the vibrations. If they don't damp out quickly the shock absorbers need to be replaced. A certain amount of friction is necessary for a smooth ride. How can we find the right amount for the most effective damping?

1. Simplifying the System

The second-order linear equation

$$m\frac{d^2x}{dt^2} + c\frac{dx}{dt} + kx = 0 \tag{1}$$

governs the motion of a weight of mass m attached to a spring of stiffness k, damped by a viscous frictional force of strength c. The variable $x(t)$ describes the mass's displacement from its equilibrium position.

Our first goal is to visualize the behavior of this system for different values of the parameters m, k, and c. With three separate parameters to vary, you might worry that there could be an enormous number of different possibilities to consider, but miraculously, a *single* combination of these parameters tells you essentially how the system will behave—that's the point of the following question.

1.1 Show that if we define a new time variable $\tau = \omega_0 t$, where $\omega_0 = \sqrt{k/m}$ is the natural frequency of the undamped system, then Equation (1) simplifies to $\frac{d^2x}{d\tau^2} + 2b\frac{dx}{d\tau} + x = 0$. Give a formula for $2b$ in terms of m, k, and c. (The factor of 2 in front of b probably looks strange, but we'll see later that it simplifies the formula for the eigenvalues.)

The advantage of the new variables is that we have only one parameter to vary, namely b. In other words, we can set $m = 1$ and $k = 1$ without loss of generality! Therefore, from now on we can restrict our attention to the simpler equation

$$\frac{d^2x}{dt^2} + 2b\frac{dx}{dt} + x = 0 . \tag{2}$$

Note that we have also reverted to the familiar letter t instead of τ.

2. Eigenvalues

2.1   Show that Equation (2) has a solution of the form $x = e^{\lambda t}$, and find and solve the characteristic equation for the eigenvalue λ in terms of b. Show that the eigenvalues are real and negative for $b > 1$ (*overdamped* case), and complex with Re $\lambda < 0$ for $0 < b < 1$ (*underdamped* case). Show that there is a repeated eigenvalue for $b = 1$ (*critically damped* case). What is special about the eigenvalues when $b = 0$ (*undamped* case)?

Open the **Damped Vibrations** tool. Its purpose is to help you gain some intuition about the meaning of undamped, underdamped, critically damped, and overdamped vibrations. The slider allows you to set the parameter b.

2.2   Move the slider all the way to the left, so that $b = 0$. This is the familiar case of a simple harmonic oscillator without damping. The eigenvalues are $\lambda = \pm i$, as shown in the display. These eigenvalues are also plotted as dots in the top panel, which shows their location in the complex λ plane.

Describe how the eigenvalues change as you drag the slider to the right.

2.3   As you drag the slider, notice that the eigenvalues move around in the complex λ plane. The picture suggests that the eigenvalues lie on the unit circle $|\lambda| = 1$ for the underdamped case $0 < b < 1$. Prove this.

3. Time Series and Animations

Besides showing the eigenvalues as a function of b, the **Damped Vibrations** tool also shows the solution $x(t)$ of Equation (2), starting from an initial condition $x(0) = 1$, $\dot{x}(0) = 0$. The corresponding motion of the mass is shown in the animation alongside the graph of $x(t)$. Notice that the animation and the time series have the same vertical scale: the current position of the mass is shown as a moving yellow dot on the graph of $x(t)$. The gray curves show the "envelope"—the time series stays between these curves.

3.1   Find the solution $x(t)$ for the given initial conditions for the underdamped case.

3.2   Move the slider toward $b = 1$. How do the vibrations change as b approaches 1 from below?

3.3   What is the qualitative difference between the solutions for $0 \le b < 1$ and $b \ge 1$?

4. What's Special about Critical Damping?

Open the **Critical Damping** tool. It allows you to compare the solutions of Equation (2) for different values of b. All the solutions start from the same initial condition, $x(0) = 1$, $\dot{x}(0) = 0$. The reference curve shown on the graph is the solution for the critically damped case, which has a special property, as you're going to see.

4.1   Move the slider for b. Notice how the solutions for different b compare to the critically damped case. By experimenting, find the value of b such that the solution $x(t)$ gets small and *stays small* as rapidly as possible.

4.2   Give a mathematical explanation of the previous answer. Why does that value of b produce the fastest decay?

Lab 10: Tool Instructions

Damped Vibrations Tool

Setting Initial Conditions

Click the **[Start]** button to start a trajectory using preset initial conditions.

Clicking in the time series will set an initial value of x and start a trajectory.

Clicking in the plane while a trajectory is being drawn will start a new trajectory.

Parameter Slider

Use the slider to set the damping coefficient b.

Press the mouse down on the slider knob, and drag the mouse back and forth to change it, or click the mouse in the slider channel at the desired value for the parameter.

Buttons

Click the **[Pause]** button to stop a trajectory without canceling it.

Click the **[Continue]** button to resume the motion of a paused trajectory.

Critical Damping Tool

Parameter Slider

Use the slider to set the damping coefficient b. The slider controls the shape of the blue or red damping trajectory.

Press the mouse down for the parameter you want to change, and drag the mouse back and forth to change it, or click the mouse in the slider channel at the desired value for the parameter.

Time Series Buttons

The buttons labeled

[ ] x

[ ] 100x

toggle the magnification level for viewing the damping behavior.

Forced Vibrations: An Introduction

11

Tools Used in Lab 11
Damped Forced Vibrations

The addition of a sinusoidal forcing function to the model of a linear oscillator brings a new level of complexity. Transient responses may appear and fade and the long term system response is closely tied to the frequency of the forcing function.

1. The Damped Oscillator with Forcing

In this lab we explore the effect of a sinusoidal forcing function on the damped and undamped mass-spring system. This lab presents an overview. For a more detailed exploration, look at Lab 10, Free Vibrations and Lab 12, Forced Vibrations: Advanced Topics, and their corresponding tools.

The equation for the damped linear oscillator with forcing function $F_0 \cos(\omega t)$, where F_0 and ω are constants, is given by

$$m\ddot{x} + b\dot{x} + cx = F_0 \cos(\omega t) \qquad (1)$$

where $b > 0$ and $F_0,\ \omega \neq 0$.

Open the **Damped Forced Vibrations** tool. Note that the vertical position as a point on the time series graph corresponds to the vertical position of the mass. You can set the initial position x_0 and velocity v_0 on the phase plane.

1.1   Set $m = 1.5$, $b = 0.5$, $k = 1$, and $F_0 = 2$, $\omega = 1$. Start the graph by setting the initial conditions on the phase plane. Sketch the result below with the three kinds of curves that make up the x vs. t graph carefully labeled and distinguished by colored pencil or dotted and dashed lines.

The next few exercises are to help you understand what is going on in this graph.

The position function $x(t) = x_{transient} + x_{steady\,state}$ is the solution of Equation (1), where $x_{transient} = x_h$ is the solution to the associated homogeneous equation and $x_{steady\,state} = x_p$ is the particular solution.

The **transient solution** is the solution of the associated homogenous equation $m\ddot{x} + b\dot{x} + kx = 0$. In the previous lab, we learned that solving the characteristic equation $m\lambda^2 + b\lambda + k = 0$ using the quadratic formula gives rise to three distinct situations:

- the **overdamped case** when $b^2 - 4mk > 0$:

$$x_h = c_1 e^{\lambda_1 t} + c_2 e^{\lambda_2 t} \text{ where } \lambda_1, \lambda_2 = \frac{-b \pm \sqrt{b^2 - 4mk}}{2m} \text{ are both negative.}$$

- the **critically damped** case when $b^2 - 4mk = 0$:

$$x_h = c_1 e^{\lambda t} + c_2 t e^{\lambda t} \text{ where } \lambda = -\frac{b}{2m}$$

- the **underdamped** case when $b^2 - 4mk < 0$:

$$x_h = e^{-\alpha t}\left(c_1 \cos(\beta t) + c_2 \sin(\beta t)\right) \text{ where } \lambda_1, \lambda_2 = \alpha \pm \beta i, \ \alpha = -\frac{b}{2m}, \ \beta = \frac{\sqrt{4mk - b^2}}{2m}$$

These solutions are called **transient** because each solution tends toward zero as t gets large.

1.2 How do the above expressions guarantee that the homogeneous solutions x_h die out with increasing time? Give a concise explanation for each of the three cases. Note that the critically damped case may require l'Hôpital's Rule.

a. overdamped:

b. critically damped:

c. underdamped:

1.3   a.   In Exercise **1.1**, did the transient solution correspond to the overdamped, critically damped, or underdamped case?

      b.   Now vary the slider on the damping constant b, holding m and k constant, to obtain each type of transient behavior. State your values for k and m and give the corresponding values of b for each type of behavior.

The Steady-State Solution

The first thing to observe is that the frequency of the steady-state solution is exactly that of the external force. Note that the method of undetermined coefficients can be used to find the steady-state solution. Start by varying the frequency of the forcing function to see how the frequency of the steady-state solution is affected. Then vary the amplitude F_0 of the forcing function. Note that the scale on the vertical axis is changing as you change ω.

1.4   a.   True or false: The steady-state solution has the same amplitude as the driving force.

      b.   True or false: The steady-state solution may be out of phase with the driving force. (*Hint:* Look at the relative motion of the orange piston and the yellow weight.)

2. The Forced Undamped System

Set $b = 0$ for the remainder of this lab. Equation (1) then becomes:

$$m\ddot{x} + kx = F_0 \cos(\omega t) \qquad\qquad (2)$$

The position function $x(t)$ is still the sum of the homogeneous and the particular solutions, but they are no longer designated as transient or steady-state. In fact the homogeneous solution is no longer transient in that it doesn't die out with increasing time and the particular solution may no longer be at all "steady." The particular solution can oscillate or grow in amplitude as time increases, depending on the frequency of the forcing function.

Recall that the homogeneous solution is $x_h = c_1 \cos(\omega_0 t) + c_2 \sin(\omega_0 t)$ where the natural frequency of the spring is $\omega_0 = \sqrt{k/m}$. Notice that x_h does not die out with time.

2.1. Assume $\omega \neq \omega_0$. Note that we can find x_p, the particular solution for Equation (2), by using the method of undetermined coefficients. As your textbook will show, x_h is a sinusoidal function of frequency ω_0, and x_p is a sinusoidal function of frequency ω. Use the software to observe that the position function x is the sum of these two sinusoidal functions of different frequencies. Using three colors or dotted and dashed lines, sketch and clearly label the graphs of these functions below.

x

t

2.2 Now set $m = 1$, $k = 1$, and $\omega = 0.5$ rad/sec. Let the values of ω approach ω_0. For what values of ω do beats begin to appear? What happens as ω gets closer and closer to ω_0? Describe what happens to the beats. Do they have a greater amplitude? Is their frequency increasing or decreasing? Remember that the scale on the vertical axis is changing as you change ω.

2.3 Now we get to the exciting part! Set $\omega = \omega_0 = 1$. The effects you observe are called **resonance.** Rather than expanding the scale to accommodate vibrations of huge amplitude, the screen shows the consequences of a "real" resonance effect. To see what happened mathematically, we must solve Equation (2) using the method of undetermined coefficients. Now $x_p = A\,t\cos(\omega_0 t) + B\,t\sin(\omega_0 t)$ for some appropriate A and B. As your textbook will show, this process yields the solution $x_p = \dfrac{F_0}{2m\omega_0} t\sin(\omega_0 t)$. Sketch a graph of the particular solution below. What happens to the amplitude of the oscillations as time increases?

Lab 11: Tool Instructions

Damped Forced Vibrations Tool

Setting Initial Conditions

Click the [Start] button to start a trajectory using preset initial conditions.

Clicking in the time series will set an initial value of x and start a trajectory.

Clicking in the plane while a trajectory is being drawn will start a new trajectory.

Parameter Sliders

Use the slider to change the values for the parameters m, b, k, F, and ω.

Press the mouse down on the slider knob for the parameter you want to change and drag the mouse back and forth, or click the mouse in the slider channel at the desired value for the parameter.

Time Series Buttons

The buttons labeled

[ ] **transient**

[ ] **steady state**

[ ] **position**

toggle the time series graphs on and off.

Other Buttons

Click the [Pause] button to stop a trajectory without canceling it.

Click the [Continue] button to resume the motion of a paused trajectory.

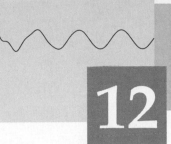

Forced Vibrations: Advanced Topics

12

Tools Used in Lab 12
   Vibrations: Amplitude
      Response
   Vibrations: Phase Response
   Vibrations: Input/Output

How does a damped harmonic oscillator behave when it is driven by a sinusoidal force? The answer is relevant to the structural vibrations of bridges and skyscrapers in a gusty wind, the tuning of a radio, a car driving over a bumpy road, and many other systems subjected to periodic excitations.

1. Forced Damped Oscillator Response

The second-order linear equation

$$m\frac{d^2x}{dt^2} + c\frac{dx}{dt} + kx = F_0 \cos \omega t \tag{1}$$

governs the motion of a damped harmonic oscillator driven by a sinusoidal force. As before in Lab 10, **Free Vibrations**, we suppose that a weight of mass m is attached to a spring of stiffness k and damped by a viscous frictional force of strength c. The variable $x(t)$ describes the displacement of the mass. The new feature is that the system is driven by the periodic force $F_0 \cos \omega t$, where F_0 is the driving strength and ω is the driving frequency.

In Lab 10, we saw that if there is no forcing ($F_0 = 0$), the motion eventually damps out: $x(t) \to 0$ as $t \to \infty$. But if $F_0 \neq 0$, the behavior becomes much more interesting, as you'll see below. The applied forcing can counteract the effects of damping. In particular, if the system is forced at frequencies close to its natural frequency, and if the damping is not too large, then "resonance" occurs, thereby causing the oscillations in $x(t)$ to grow to large amplitudes. But what happens if the system is jiggled much faster than its natural frequency, or much slower? The goal of this lab is to help you understand the different types of responses—both resonant and non-resonant—that the system can exhibit.

Equation (1) has five parameters: m, k, c, F_0, ω. This is a lot of parameters to vary. To keep things simple, we assume that $m = 1$, $k = 1$, and $F_0 = 1$. This might seem like an overly special case, but in fact it is completely general, in the following sense: if m, k, and F_0 are all nonzero, we can always convert Equation (1) to an equation of the form

$$\frac{d^2x}{dt^2} + 2b\frac{dx}{dt} + x = \cos \omega t \tag{2}$$

by rescaling time and x appropriately. (For now, take that on faith. If you don't believe that anything so wonderful could possibly be true, see the last question at the end of this lab.) The upshot is that we can get away with studying Equation (2), which has only two parameters instead of five, and yet we aren't sacrificing any generality!

1.1 Show that Equation (2) has a particular solution (known as the **forced response**) given by

$$x_p(t) = A\cos(\omega t - \phi) \tag{3}$$

where the **amplification factor** A is given by

$$A = \frac{1}{\sqrt{(1-\omega^2)^2 + 4b^2\omega^2}} \tag{4}$$

and the **phase lag** ϕ is

$$\phi = \tan^{-1}\left(\frac{2b\omega}{1-\omega^2}\right). \tag{5}$$

2. Amplitude Response

Our goal in this part is to understand the meaning of Equation (4).

Open the **Vibrations: Amplitude Response** tool. The schematic shows a mass on a spring, along with a dashpot that damps the oscillations of the mass. The whole system is being driven by a piston that moves the top of the spring up and down. (More intuitively, imagine holding the top of the spring in your hand, and jiggling it up and down periodically.) The time series graph shows $x(t)$, the solution of Equation (2), for the initial conditions $x(0) = 0$, $\dot{x}(0) = 0$, and for the values of b and ω chosen on the sliders. In the upper-right part of the screen, the curve (4) is plotted as a function of ω, for a given value of b. This graph of A vs. ω is known as the **amplitude response curve**.

2.1 Set $b = 0.25$. Then move the slider for ω back and forth, and notice that a horizontal line is being drawn on the time series graph at a height given by A. Now choose $\omega = 1$. Click on the **[Start]** button to see the resulting solution $x(t)$. How is the eventual amplitude of the oscillations in $x(t)$ related to A? Explain your observations, using the concepts of homogeneous and particular solutions.

2.2   For $b = 0.25$ and $\omega = 1$, what is the value of A predicted by (4)? Does this agree with what you observe in $x(t)$?

2.3   Increase the driving frequency to $\omega = 2$. How does $x(t)$ change, compared to the previous results?

So far you have held the damping strength b fixed at $b = 0.25$. Now explore what happens when you vary b.

2.4   As you decrease b below 0.25, describe what happens to the shape of the amplitude response curve. In particular, what happens to the height of the peak in the curve? Also, roughly estimate the value of ω at which the peak occurs.

2.5   Now suppose you increase b. What happens to the height and location of the peak in the curve?

2.6   Perhaps you have noticed that the peak in the amplitude response curve occurs near $\omega - 1$ when b is small, but shifts to the left for larger b. By maximizing Equation (4) with respect to ω, find a formula for the value of ω at which the peak occurs.

2.7   Find the height of the peak, as a function of b. This quantity is known as Q, the **quality factor** of the system. Give an approximate formula for Q if b is small.

Conclusion

We have seen that if the damping b is small, the amplification factor A becomes large when $\omega \approx 1$. This is the phenomenon of **resonance**: if a vibrating system is weakly damped, it exhibits a large amplification factor when driven at a frequency near its natural frequency (which is $\omega = 1$ because of the way we have scaled the parameters). The strength of the resonance is commonly expressed in terms of the peak amplification factor, that is, the quality factor Q of the system. For instance, electrical engineers try to design radio amplifiers with a high Q, to allow for precise tuning and strong amplification of faint signals.

3. Phase Response

In this part, we explore the phase relationship between the oscillating mass and piston. Do they move together, or in opposition, or what? How does the answer depend on ω and b?

3.1   Open the **Vibrations: Amplitude Response** tool again. Set $b = 0.15$. When $\omega = 2$, does the mass eventually move **in phase** with the piston (do they move up together, and down together), or does the mass move in **antiphase** (up when the piston moves down, and vice versa)?

3.2   Repeat the previous question, for $\omega = 0.67$. After the system settles down, does the mass move in phase with the piston, or in antiphase?

3.3   Give a rule of thumb that generalizes the previous results. For $b = 0.15$, when do antiphase oscillations occur, and when do in-phase oscillations occur?

Now open the **Vibrations: Phase Response** tool to explore these issues in more detail. The graph of ϕ vs. ω is called the **phase response curve**. It is a graph of Equation (5) as a function of ω, for a given value of b. The meaning of phase lag ϕ becomes clear if you look at the time series of the forcing $F(t) = \cos \omega t$, and compare it to the response $x(t)$, which eventually approaches $x_p(t) = A \cos(\omega t - \phi)$ after transients decay. The predicted phase lag ϕ is indicated by the green arrow. When $\phi \approx \pi$, the two curves are about half a cycle apart, and hence are in antiphase, whereas when $\phi \approx 0$, they are in phase.

3.4   The **Vibrations: Phase Response** tool includes sliders for both ω and b. Play with the b slider to see how it affects the phase response curve. How does the shape of the curve change as b is varied?

3.5   Using Equation (5), what are the predicted values of ϕ as $\omega \to 0$, $\omega \to 1$, and $\omega \to \infty$?

4. Input vs. Output

Another way to visualize the effects of the periodic drive is to plot the solution $x(t)$ vs. the applied forcing $F(t)$. Such a graph is called an **input-output graph**. The idea is that the input $F(t)$ is converted or "filtered" by the system to produce the output $x(t)$. For instance, in electrical circuits (Lab 13), $F(t)$ might represent an applied AC voltage, and $x(t)$ would represent the output AC current. The shape of the resulting input-output graph provides a great deal of information about the amplitude and phase response of the system.

Open the **Vibrations: Input/Output** tool. As before, Equation (2) is integrated starting from $x(0) = 0$, $\dot{x}(0) = 0$ and for the values of b and ω chosen on the sliders. The left panel shows the input-output graph in the (F, x) plane. The transient part of the solution is shown in gray, and the long-term forced response is shown in red. The right panel shows the time series of the input $F(t) = \cos \omega t$, along with the time series of the output $x(t)$.

Behavior Near Resonance

Throughout this part, set $\omega = 1$.

4.1   Choose a value of b and look at the input-output graph. Prove that the curve approaches an ellipse, and find the Cartesian equations of this limiting ellipse.

4.2   Click on the **[Start]** button to plot the corresponding time series of $x(t)$. Observe that $x(t)$ crosses through 0 at the precise instant when $F(t)$ is a maximum or a minimum. Explain this mathematically.

4.3   By estimating from the graph, what is the height of the ellipse when $b = 0.25$? $b = 0.50$? $b = 1$? Guess a formula for the height of the ellipse as a function of b, and prove this formula if you can. Explain how all this is related to the quality factor Q discussed earlier.

Weak Damping

Throughout this part, set $b = 0.25$.

4.4   What happens to the ellipse for $\omega > 1$? $\omega < 1$?

4.5   What do you think happens to the ellipse in the limit $\omega \to 0$? $\omega \to \infty$? Try to justify your guesses mathematically.

4.6   Set $\omega = 0.25$. Click on the **[Start]** button, wait for the times series to be drawn, and then click on the **[Continue]** button. Explain why the two time series are almost the same. And why is the ellipse confined almost entirely to the $45°$ diagonal line?

5. Scaling the System

5.1　Show that if m, k, and F_0 are all nonzero, Equation (1) can be rescaled to Equation (2) by introducing new dimensionless definitions of x and t, and defining b appropriately.

Lab 12: Tool Instructions

Vibrations: Amplitude Response Tool

Setting Initial Conditions

Click the [**Start**] button to start a trajectory using preset initial conditions.

Clicking in the time series will set an initial value of x and start a trajectory.

Clicking in the plane while a trajectory is being drawn will start a new trajectory.

Parameter Sliders

Use the slider to change the values for the parameters b and ω.

Press the mouse down on the slider knob for the parameter you want to change and drag the mouse back and forth, or click the mouse in the slider channel at the desired value for the parameter.

Buttons

Click the [**Pause**] button to stop a trajectory without canceling it.

Click the [**Continue**] button to resume the motion of a paused trajectory.

Vibrations: Phase Response Tool

Setting Initial Conditions

Click the [**Start**] button to start a trajectory using preset initial conditions.

Clicking in the time series will set an initial value of x and start a trajectory.

Clicking in the plane while a trajectory is being drawn will start a new trajectory.

Parameter Sliders

Use the slider to change the values for the parameters b and ω.

Press the mouse down on the slider knob for the parameter you want to change and drag the mouse back and forth, or click the mouse in the slider channel at the desired value for the parameter.

Buttons

Click the [**Pause**] button to stop a trajectory without canceling it.

Click the [**Continue**] button to resume the motion of a paused trajectory.

Vibrations: Input/Output Tool

Setting Initial Conditions

Click the [**Start**] button to start a trajectory using preset initial conditions.

Clicking in the time series will set an initial value of $x(t)$ and start a trajectory.

Clicking in the plane while a trajectory is being drawn will start a new trajectory.

Parameter Sliders

Use the slider to change the values for the parameters b and ω.

Press the mouse down on the slider knob for the parameter you want to change and drag the mouse back and forth, or click the mouse in the slider channel at the desired value for the parameter.

Buttons

Click the [**Pause**] button to stop a trajectory without canceling it.

Click the [**Continue**] button to resume the motion of a paused trajectory.

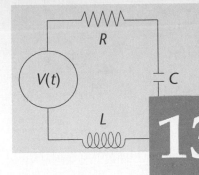

13 Electrical Circuits

Tools Used in Lab 13
Series Circuits
Damped Vibrations: Energy
Van der Pol Circuit

A series circuit with an inductor, resistor, and capacitor can be represented by $L\ddot{q} + R\dot{q} + \dfrac{1}{C}q = V(t)$, a second-order linear differential equation with constant coefficients. Look familiar?

1. Circuit Laws

We know the following two facts about our series circuit (due to Kirchhoff):

1. The current in every part of the circuit is the same.
2. The sum of the voltage drops around the circuit must be equal to the input voltage $V(t)$.

From these facts, we can obtain a differential equation where each term on the left represents a voltage drop (in volts) across the designated circuit element and each constant L, R, and $\dfrac{1}{C}$, can be viewed as a constant of proportionality in henries, ohms, or (farads)$^{-1}$, respectively. Note that q represents the charge (on the capacitor) as a function of time t and $I = \dfrac{dq}{dt}$ represents current in the circuit as a function of time t. The equation can have three forms, based on convenience. The **basic series circuit equation** is

$$L\dot{I} + RI + \frac{1}{C}q = V(t) \tag{1}$$

In terms of charge, $q(t)$, in coulombs (using the fact that $I = \dfrac{dq}{dt}$):

$$L\ddot{q} + R\dot{q} + \frac{1}{C}q = V(t) \tag{2}$$

In terms of current, $I(t)$, in amperes (by differentiating every term of the basic equation):

$$L\ddot{I} + R\dot{I} + \frac{1}{C}I = \dot{V}(t) \qquad\qquad (3)$$

For the most part, we use Equation (2) in terms of $q(t)$. Note that it looks surprisingly like the equation for the spring:

$$m\ddot{x} + b\dot{x} + kx = F(t)$$

In fact, within the linear operation of the circuit elements, the analogy is very tight and quite useful in understanding the behavior of circuits.

2. L-C Circuits

2.1   The L-C Circuit: $L\ddot{q} + \frac{1}{C}q = 0$

Open the **Series Circuits** tool. Set $R = 0$ and $A = 0$ *precisely*. Without an input voltage, the only energy in the circuit is due to the initial conditions. Start with (nontrivial) initial conditions by clicking the mouse on the phase plane to set $q(0) = q_0$ and $q'(0) \equiv I(0) = I_0$. Are you surprised to get the equivalent of simple harmonic motion on a spring?

a.   Solve the linear differential equation for $q(t)$ above in terms of L and C.

b.   What is the natural frequency of oscillation for the circuit? Compare it to the natural frequency of the spring, $\omega_0 = \sqrt{k/m}$.

The capacitor stores electrical energy and has a voltage drop in proportion to its charge, so that $1/C$ *is analogous to k, the spring constant.* The inductor stores magnetic energy and builds a 'back' voltage in proportion to the change in the current flowing through it, dI/dt, so *the inductance L is analogous to the mass m,* (i.e. the inertia or resistance to change in velocity) in the spring system.

Notation: for convenience, we use "$\approx$" for "is analogous to," so that $\begin{aligned} L &\approx m \\ \frac{1}{C} &\approx k \end{aligned}$

2.2   Energy in the L-C Circuit

Look at the energy graph on the **Damped Vibrations: Energy** tool. Use the analogy with the mass-spring system to obtain the expression of the magnetic (kinetic) energy in the inductor and the electric (potential) energy in the capacitor. Notice that without resistance there is no dissipation of energy in the circuit, so the total energy remains constant. Remember that $q \approx x$, the displacement, and $I \approx v$, the velocity $\dot{x}$. Complete the sentence for the energy in the circuit:

mass-spring $\qquad E_{total} = E_{kinetic} + E_{potential} = \dfrac{1}{2}mv^2 + \dfrac{1}{2}kx^2$

L-C circuit $\qquad E_{total} = E_{magnetic} + E_{electric} = $ _____

2.3   The Forced L-C Circuit: $L\ddot{q} + \dfrac{1}{C}q = A\cos(\omega t)$

Use the **Series Circuits** tool to determine the response when the frequency of the input voltage $V(t) = A\cos(\omega t)$ is near or equal to the natural oscillating frequency ω_0 of the circuit. Make a sketch of the resulting oscillations when ω is close to but unequal to ω_0.

3. L-R-C Circuits

3.1 The L-R-C Circuit with $V(t) \equiv 0$: $L\ddot{q} + R\dot{q} + \dfrac{1}{C}q = 0$

a.   What does the resistance, R, in a series circuit correspond to in the spring system?

b.   Now set $A = 0$ and set R to a small nonzero value on the sliding scale. What kind of motion do you observe? Give a rough sketch below. State your initial conditions for the charge q_0 and current $\dot{q}(0) = I(0)$.

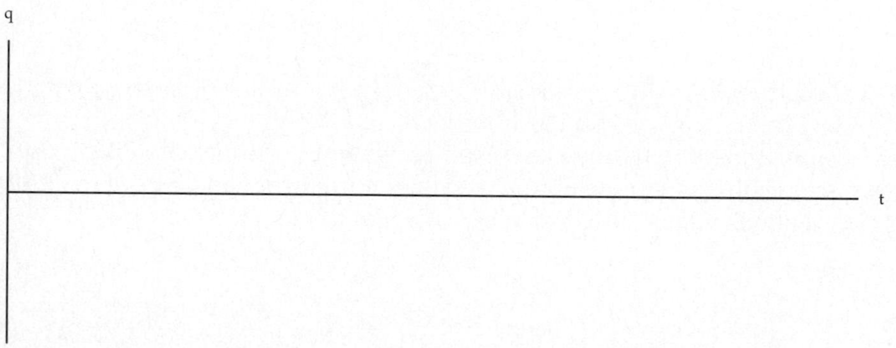

c.   Are the charge q and current I underdamped, overdamped, or critically damped?

d.   Using the analogy with the spring for the critical damping, $b_{cr} = \sqrt{4mk}$, find the value of the resistance R_{cr} that gives critical damping.

e.   Now use $L\ddot{q} + R\dot{q} + \dfrac{1}{C}q = 0$, via the characteristic equation $L\lambda^2 + R\lambda + \dfrac{1}{C} = 0$, to find again the resistance R_{cr} that gives critical damping.

f.   We purposely left the sliders without units so that any consistent set of realistic units could be used. Set $L = 2$ henries and $C = 0.5$ microfarads. Vary the resistance to find the resistance that gives critical damping. With the units we are using, the slider for the resistance is in kilo-ohms (that is, 10^3 ohms) and the time scale is in milliseconds (10^{-3} seconds). Is the critical resistance you discovered the same as the calculated resistance R_{cr}? If not, figure out the problem. It should be the same. What is the resistance ?

g.   For every value of $R \neq 0$, what is the long-term behavior of $q(t)$ and $I(t)$?

3.2   Energy in the L-R-C Circuit

Now that we have resistance in the circuit, there is heat loss. In fact, the loss due to heat dissipated in the resistor is $E_{dissipated} = E_{total} - (E_{magnetic} + E_{electric})$. Keep the analogy with the mass-spring system firmly in mind. Describe carefully what happens over the long run to the available energy $E_{magnetic} + E_{electric}$ (or in the case of the spring, $E_{potential} + E_{kinetic}$). What happens to the energy that is dissipated?

3.3   The L-R-C Circuit with $V(t) \equiv A\cos(\omega t)$: $L\ddot{q} + R\dot{q} + \dfrac{1}{C}q = A\cos(\omega t)$

Reopen the **Series Circuits** tool. Set R to some nonzero values and experiment with the response of the circuit to various values of ω and $A \neq 0$.

a.   Describe what happens on the phase plane and on the time series as t becomes large.

b.   The transient solutions for the charge (and current) are analogous to the transient solutions for the position (and velocity) in Lab 11, Forced Vibrations: An Introduction and die out as time becomes large. Was that your experience in Exercise **3.1(g)**?

c.   The steady-state solution depends on the impressed voltage as well as the circuit elements and gives the long-term behavior of the system. **Set $A = 2$ and $\omega = 1$. Set $C = 0.5$ microfarads $R = 1$ ohm, and $L = 1$ henry.** Now try several values for L, $1/C$, and ω. Note that the wiggle at the beginning of the solution curve is due to the transient behavior of the circuit. What is the angular frequency of the steady-state solution?

d.   The transient solution corresponds to the _____ solution (homogeneous or particular?) of the differential equation for the circuit with applied voltage.

The steady state solution corresponds to the _____ solution (homogeneous or particular?).

4. Additional Exercises

4.1   A Simplified Induction Motor

Suppose we have a new induction motor circuit with a problem. The AC power supply overheated at small frequencies, so we added a capacitor of 1 microfarad in series, but now the motor circuit has undesirable resonances as we accelerate it through its natural frequency. There are already 500 ohms of resistance in the circuit, but it has been suggested that an additional resistor be placed in series to damp out the resonance. If the inductance L is 1 henry, find the least resistance required to damp out the resonance in the circuit.

4.2   A Nonlinear Example

Just for fun, open the **Van der Pol Circuit** tool . The resistor is nonlinear and supplies variable damping, that is, positive damping (which dissipates energy) for part of the time and negative damping (which adds energy) for part of the time.

Look for a limit cycle! (That is, try a variety of initial conditions. What happens?) You don't have to look very hard, because the limit cycle is a "global attractor." Describe what happens.

Lab 13: Tool Instructions

Series Circuits Tool

Setting Initial Conditions
Click the mouse on the $q\dot{q}$ plane or the time series graph to set the initial conditions for a trajectory.

Clicking in the $q\dot{q}$ plane or the time series graph while a trajectory is being drawn will start a new trajectory.

Parameter Sliders
Use the sliders to change the values for the parameters L, R, $1/C$, A, and ω. The circuit is automatically redrawn when the slider values are set to zero.

Press the mouse down on the slider knob for the parameter you want to change and drag the mouse back and forth, or click the mouse in the slider channel at the desired value for the parameter.

Time Series Buttons
The buttons labeled
   [ ] **charge**
   [ ] **current**
toggle the time series graphs on and off.

Other Buttons
Click the [**Pause**] button to stop a trajectory without canceling it.
Click the [**Continue**] button to resume the motion of a paused trajectory.
Click the mouse on the [**Clear**] button to remove all the trajectories from the graph.

Damped Vibrations: Energy Tool

Setting Initial Conditions
Click the [**Start**] button to start a trajectory using preset initial conditions.
Clicking in the time series graph will set an initial value of x and start a trajectory.
Clicking in the time series graph while a trajectory is being drawn will start a new trajectory.

Parameter Slider
Use the slider to set the damping coefficient, b.
Press the mouse down on the slider knob for the parameter you want to change and drag the mouse back and forth, or click the mouse in the slider channel at the desired value for the parameter.

Time Series Buttons
The buttons labeled
   [ ] **energy**
   [ ] **velocity**
   [ ] **Position**
toggle the time series graphs on and off.

Other Buttons
Click the [**Pause**] button to stop a trajectory without canceling it.
Click the [**Continue**] button to resume the motion of a paused trajectory.

Van der Pol Circuit Tool

Setting Initial Conditions

Click the mouse on the *iv* graphing plane to set the initial conditions for a trajectory.

Clicking in the *iv* plane while a trajectory is being drawn will start a new trajectory.

Parameter Sliders

Use the slider to set the nonlinear resistance parameter e.

Press the mouse down on the slider knob for the parameter you want to change and drag the mouse back and forth, or click the mouse in the slider channel at the desired value for the parameter.

Time Series Buttons

The buttons labeled

   [ ] **capacitor voltage**

   [ ] **current**

   [ ] **nonlinear resistance**

toggle the time series graphs on and off.

Other Buttons

Click the mouse on the **[Clear]** button to remove all the trajectories from the graph.

"Beam me over, Laplace!"

14 Laplace Transforms

Where do linear differential equations act like algebraic equations? In the space of Laplace transforms.

1. Definition and Properties

Definition of the Laplace Transform

For a function f defined on $[0,\infty)$, the Laplace transform is the function of s defined by

$$\mathbf{L}\{f(t)\}(s) = F(s) = \int_0^\infty e^{-st} f(t)\, dt \qquad (1)$$

for all s for which the integral converges. We say that $F(s)$ is now a function in the **s-domain**.

1.1 In order to understand the concepts that are part of this definition, open the **Laplace: Definition** tool. This tool illustrates the case when $f(t) = \sin(t)$, but this simple example can lead us to a more general understanding of the definition for any function f. The following questions will lead the way. Begin by considering the upper-half of the screen.

   a. For a positive value of s, what does multiplication by e^{-st} do to the function f?

   b. What happens to $e^{-st}f(t)$ as you increase the value of s?

c. As you may recall from calculus, the integral (1) with an infinite limit is called an **improper integral**. This integral is defined to be the $\lim\limits_{T\to\infty}\int_0^T e^{-st}f(t)dt$ and is said to **converge** whenever the (finite) limit exists. In the lower-left part of the screen, we look at an intermediate step where we plot the function $I(s,T) = \int_0^T e^{-st}f(t)dt$ versus T. (Note that you can select a value for T by using the cursor on either graphing plane on the left of the screen.) That is, we evaluate the definite integral from 0 to T. Describe in words what this integral represents.

d. For a given value of s, what happens to $I(s,T)$ as T becomes large? Does that happen eventually for every value of $s > 0$? Can you visually determine the $\lim\limits_{T\to\infty}\int_0^T e^{-st}f(t)dt$ for each value of s?

e. Now put the cursor on the s-slider and slide s from 0 to the end. Describe what happens to all three graphs as s increases.

    i. the graph of $e^{-st}f(t)$ versus t:

    ii the graph of $I(s,T)$ versus t:

    iii. the graph of $F(s)$ versus s:

1.2   What kinds of functions on $[0, \infty)$ have Laplace transforms? Not every function does, but we can guarantee the existence of a Laplace transform for any function f on $[0, \infty)$ that is piecewise continuous (so it can be integrated) and is of **exponential order** α. The latter condition means that the function f must not grow too fast as t increases, in that $|f(t)|$ must be bounded eventually by some exponential function $Me^{\alpha t}$ for some $M > 0$. Note that it is precisely this property (that $f(t) = \sin(t)$ was "dampable" by multiplication by e^{-st} for some $s > 0$) that plays such a significant role in the **Laplace: Definition** tool.

   a.   Not every function has an exponential order. Can you think of one that does not have that property?

Also note that there may be functions out there with neither exponential order nor piecewise continuity that have Laplace transforms. Later, we'll meet the Dirac delta function which has neither property and, in fact, is not strictly a function, yet it has a useful Laplace transform.

1.3   Open up the **Laplace: Transformer** tool. Send a few standard functions through the "transformer" by clicking on the arrow to see what they transform to in the s-domain. Note that every function in the list meets the two criteria in the preceeding paragraph. What do you observe about the transformed function $F(s)$ as s becomes large?

Because Laplace transforms are just integrals, they share the property with derivatives and integrals of being linear operators. This means that the **linearity property** $L\{a\,f + b\,g\} = aL\{f\} + bL\{g\}$ holds for any functions f and g whose Laplace transforms exist in a common domain, and any constants a and b.

The fundamental reasons that Laplace transforms are so useful are linearity and the way they act on derivatives. These properties allow us to change linear differential equations with initial conditions (initial-value problems in the t-domain) into algebraic equations in the s-domain. Then we can solve them in the s-domain and transform them back to obtain the solutions in the t-domain.

The Laplace Transform of a Derivative

Differentiation in the t-domain is multiplication by s in the s-domain (for zero initial conditions).

The Derivative Theorem

If the functions f and f' are continuous on $[0, \infty)$ and f'' is piecewise continuous on $[0, \infty)$, and if f, f' and f'' have exponential order α, then for $s > \alpha$,

$$L\{f'\}(s) = s\,F(s) - f(0)$$
$$L\{f''\}(s) = s^2 F(s) - sf(0) - f'(0)$$

(2)

where $f(0)$ and $f'(0)$ are the values of f and f' corresponding to $t = 0$.

1.4   Open the **Laplace: Derivative** tool to investigate this property for $f(t) = \cos(2t)$. What is $f(0)$? What is $\mathbf{L}\{f'(t)\}$ according to Equation (2)? Is that what you would get if you took the derivative of $\cos(2t)$ first and then took the Laplace transform? Show this algebraically.

1.5   Illustrate the Derivative Theorem for the function $f(t) = t^2 + 2$ without using the computer.

1.6   Use Equation (2) to transform the linear initial-value problem $x'' + x = e^{-t}$, with initial conditions $x(0) = 1, x'(0) = 0$, to an algebraic equation. Consult the **Laplace: Transformer** tool for the transforms of the forcing function. Solve for $X(s)$, the Laplace transform of the solution $x(t)$. You need not go further at this point.

In order to finish solving the above initial value problem, we must make certain that we know how to transform $X(s)$ in the s-domain to the unique solution $x(t)$ in the t-domain. There is a bit of a problem. Because the integral in definition (1) works for piecewise continuous functions, there may be many functions with the same Laplace transform. If there is a *continuous* function with a given transform, *there can be only one!* It is this function that we choose to be the unique inverse transform. If there is no such function we can still recover a function from its transform, but there may be ambiguities at the points of discontinuity.

With these reservations, we say that the **inverse Laplace transform** of a function $F(s)$ is the function f of t that satisfies $\mathbf{L}\{f\}(s) = F(s)$. We denote it by $\mathbf{L}^{-1}\{F(s)\}$.

1.7   When you operated the **Laplace: Transformer** tool in Exercise **1.6**, did you notice that it is also an "inverse transformer"? Try the inverse transformer on the following two transforms to find the unique continuous function $f(t)$ that corresponds to each $F(s)$. Write the functions below. You will need to use the linearity property. More importantly, you will need to rewrite each term to fit the formulas provided by the **Laplace: Transformer** tool.

a. $\mathbf{L}^{-1}\{\dfrac{4}{s^3} - \dfrac{2}{s^2+1}\}$

b. $\mathbf{L}^{-1}\{\dfrac{1}{s-2} + \dfrac{5}{s} - \dfrac{2s}{s^2+9}\}$

1.8 Find the inverse transform for the function $X(s)$ determined in Exercise **1.6** in order to completely solve the given initial-value problem $x'' + x = e^{-t}$ where $x(0) = 1$, $x'(0) = 0$. Note that it is standard to use the method of partial fractions here.

Consider now the simplified unforced mass-spring problem modeled by the differential equation $x'' + 2bx' + x = 0$ with initial conditions $x(0) = 1$, $x'(0) = 0$. This physical situation was illustrated in the **Damped Vibrations** tool for Lab 10. Note that we can transform the initial-value problem using Equation (2):

Wait! Look at the denominator! Except for the label for the variable (s instead of λ), it is exactly the left-hand side of the characteristic equation $\lambda^2 + 2b\lambda + 1 = 0$ that we obtained when we solved the original initial value problem in Lab 10. Recall that the characteristic equation gave us the eigenvalue λ. The zeros of the denominator of the Laplace transform of the solution to an initial-value problem are called the *poles* of the transform. The significant fact is that the poles of the transform carry the same information as the eigenvalues of the characteristic equation.

1.9 To get these concepts clearly in mind, open the **Laplace: Vibration and Poles** tool and vary the damping constant b to obtain the four possible cases ($b = 0$, $0 < b < b_c$, $b = b_c$, $b > b_c$). Describe what happens to the poles as b is increased from 0 through 2.

1.10 Now look at the lower graph of $X(s)$ vs. s. As you move the b slider notice that the poles show up as asymptotes for the overdamped and critically damped cases. Why don't they show up for the underdamped case?

This situation is our first hint of the deeper structure of the s-domain. In more advanced courses, you will learn to treat s as a complex variable and $X(s)$ as a complex-valued function of a complex variable. However for the remainder of this lab, we will stay with the introductory treatment and consider s and $X(s)$ as real.

2. Forcing Functions

Let's look at some new forcing functions. The first theorem deals with translation along the s-axis. To get a feel for this theorem, open the **Laplace: Translation** tool, and use the slider to vary parameter a. Find both transforms and inverse transforms. This theorem is also called the First Shift Theorem.

The Translation Theorem

> If the Laplace transform of f exists for $s > \alpha$, so that $L\{f(t)\}(s) = F(s)$, then
> $$L\{e^{at}f(t)\}(s) = F(s-a) \text{ for } s > \alpha + a.$$

2.1 At this point you are ready to see the overall picture of how Laplace transforms work to solve initial-value problems. Open the **Laplace: Solver** tool and send through a few forcing functions. Sketch the solutions for the following initial-value problems with initial-values $x(0) = 1$, $x'(0) = 0$.

a. $x'' + x = e^{-t}$

b. $x'' + 3x' + 2x = \sin(t)$

As yet it is difficult to see the advantages in using Laplace transforms. We'll begin to appreciate them more as we start looking at special functions. The first one of interest is the **unit step function:**

$$u(t-a) = \begin{cases} 0 & t < a \\ 1 & t > a \end{cases}$$

2.2　Use the equation $\mathbf{L}\{u(t-a)\} = \dfrac{e^{-as}}{s}$ to obtain the Laplace transforms of the following functions. First sketch the functions, and then write them in terms of unit step functions. Then take the transforms.

a.　$f(t) = \begin{cases} 0 & t < 1 \\ 2 & 1 < t < 3 \\ -1 & 3 < t < 4 \\ 0 & t > 4 \end{cases}$

b.　the square wave; $f(t) = \begin{cases} 1 & 2n < t < 2n+1 \\ -1 & 2n+1 < t < 2n+2 \end{cases}$　where n is an integer.

c.　Express $g(t) = (1 - u(t-1))e^{-t}$ in the form of the functions in parts a. and b. and sketch a graph of the function. Then find the Laplace transform.

How do we take the Laplace transform of the product of a unit step function with another function? We now need a new theorem. This one is sometimes called the Delayed Function Theorem or, if you are counting, the Second Shift Theorem. We call will it the Shifting Theorem.

The Shifting Theorem

Suppose $F(s) = \mathbf{L}\{f\}(s)$ exists for $s > \alpha \geq 0$. If a is a positive constant,

$\mathbf{L}\{u(t-a)f(t-a)\}(s) = e^{-as}F(s)$.

An alternate form that is particularly useful is

$\mathbf{L}\{u(t-a)\,f(t)\} = e^{-as}\mathbf{L}\{f(t+u)\}$.

2.3　Open the **Laplace: Shifting Theorem** tool. Try a few values of parameter a on the slider. Take both transforms and inverse transforms by activating the arrows with the cursor. Where do "delayed functions" come in? Open the **Laplace: Shift and Step** tool while you consider this question. Then complete the following exercises.

a. You may need to review what a shifted function is. For instance, if we replace the variable f in $f(t)$ by $t - a$ where $a < 0$, the graph of the function will be shifted to the right a units. Check this result for the sine function in the tool and apply this idea to a new function $f(t) = \dfrac{t}{4}\cos(2t)$. The graph of this function before shifting is given below. Sketch the shifted function.

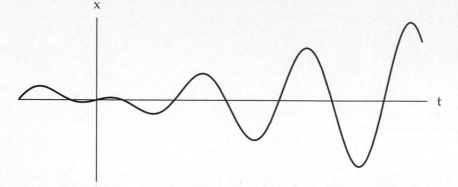

b. Describe the changes in the graph when a function $f(t)$ is multiplied by $u(t - a)$.

Explain and make a sketch using the function $f(t) = \dfrac{t}{4}\cos(2t)$.

x

t

c. Describe the new graph obtained for $u(t - a)f(t - a)$. Did we turn on the function at $t = a$ instead of $t = 0$? In other words, did we "delay" the function? Explain and make a sketch of the delayed function for $f(t) = \dfrac{t}{4}\cos(2t)$.

x

t

2.4 Suppose you found a function of s in the s-domain that was an e^{-as} multiplied by the Laplace transform of a known function. Would you suspect a delayed function? Try a few examples below by finding the inverse Laplace transform.

    a. $\mathbf{L}^{-1}\{e^{-2s}\dfrac{5}{s^3}\}$

    b. $\mathbf{L}^{-1}\{e^{-(\pi/2)s}\dfrac{s}{s^2+9}\}$

The most interesting special function is the **Dirac delta function** (which mathematicians call a **generalized function** because we have to "stretch" our concept of what a function is in order to include it). We are going to view the delta function in terms of a limiting process. First, however, let's consider the idea from physics of an **impulsive force,** a force applied over some brief time interval. Then, to calculate the **impulse** we must integrate over the time interval.

$$\text{Impulse} = \int_b^a f(t)\,dt$$

We define a new function: $f_h(t) = \dfrac{1}{h}(1 - u(t-h)) = \begin{cases} \dfrac{1}{h} & 0 < t < h \\ 0 & t > h \end{cases}$

2.5 Determine the impulse for f_h for the time interval $[0,\infty)$.

Look at the tool **Laplace: Delta Function**. Try both sliders. Note that as $h \to 0$, each "box" continues to have unit area. We are going to define the Dirac delta function to be the limit of the f_h's as $h \to \infty$ so that the property of having area 1 under the curve is maintained. For this reason, the delta function $\delta(t)$ is often referred to as the **unit impulse function**.

Are you having trouble believing in the delta function? What is the problem (that is, aside from the fact that the delta function is infinitely tall, infinitely thin, and has an area under the curve of 1)? If you are willing to accept it, you will find it very useful in modeling a variety of physical problems, especially those involving impulsive forces and voltage surges. However, it is not your everyday function. Until the development of distribution theory, some mathematicians viewed it with suspicion.

2.6   Open up the **Laplace: Solver** tool again and try some delta functions and unit step functions.

   a.   Sketch the solution for $x'' + 3x' + 2x = u(t-2)$.

   b.   Sketch the solution for $x'' + 3x' + 2x = \delta(t-2)$.

3. The Convolution Theorem and the Transfer Function

How are the solutions to a linear differential equation related to the forcing function? The answer is surprisingly easy to see in the s-domain. However, in the t-domain we need a new idea—convolution!

The **zero initial state**, where all initial conditions are zero, will be assumed for the rest of this lab. We are getting very close to the main idea!

3.1   What happens if we use $\delta(t)$, the unit impulse function at $t = 0$, as our forcing function? (Recall that $L\{\delta(t)\} = 1$.) Consider the general second-order linear differential equation with constant coefficients:

$$ax'' + bx' + cx = \delta(t)$$

For the zero initial state, show that $X(s) = \dfrac{1}{a\,s^2 + b\,s + c}$.   (3)

The right-hand side of Equation (3) is called the **transfer function G(s)**, and it represents the response of the system to the unit impulse function. It carries information that we have seen in another form. *Notice that the zeros of the denominator of G(s) are precisely the roots of the characteristic equation $a\,\lambda^2 + b\,\lambda + c = 0$.* Now we can recognize the zeros of the denominator as the poles of the transfer function G(s).

Many times the transfer function is exactly what you need to investigate a mechanical or electrical system. To find the solution in the s-domain for any forcing function for which a Laplace transform $F(s)$ exists, you need only take the product $G(s) F(s)$. Of course, if you need the solution in the t-domain, you need to find $\mathbf{L}^{-1}\{G(s)F(s)\}$. That, unfortunately, is harder than it looks!

3.2   Give an example to show that $\mathbf{L}^{-1}\{G(s)F(s)\}$ does *not* equal $g(t)f(t)$.

What then is $\mathbf{L}^{-1}\{F(s)G(s)\}$? The answer is given by the Convolution Theorem, but first we must define the convolution of two functions on f and g defined on $[0,\infty)$.

The convolution of two piecewise continuous functions f and g on $[0,\infty)$ is defined to be

$$(f * g)(t) \equiv \int_0^t f(T)g(t - T)dT, \text{ or equivalently,}$$

$$(f * g)(t) \equiv \int_0^t f(t - T)g(T)dT.$$

In order to get a feel for the convolution as the integral of the product of $f(T)$ and $g(t - T)$, open the tool **Laplace: Convolution Example.** The title of this tool is a little misleading because Laplace transforms do not play a role, as yet. The very instructive example illustrated here is due to S. Farlow and can be found in his text *An Introduction to Differential Equations and their Applications*, McGraw-Hill, Inc., 1994, p. 304.

3.3   Open the **Laplace: Convolution Theorem** tool. Select functions and activate the convolve arrow followed by the **L**-arrow; then activate the **L**-arrow followed by the multiply arrow. What do you observe about the result?

This next theorem should not surprise you!

The Convolution Theorem

Let $f(t)$ and $g(t)$ be piecewise continuous on $[0,\infty)$ and of exponential order α, then

$\mathbf{L}\{f * g\}(s) = F(s)G(s)$, or equivalently,

$\mathbf{L}^{-1}\{F(s)G(s)\} = f * g$ where $F(s) = \mathbf{L}\{f\}(s)$ and $G(s) = \mathbf{L}\{g\}(s)$.

Now we can fill in some of the missing ideas in the **Laplace: Solver** tool. The desired inverse transform of the solution in the s-domain is the convolution of $g(t)$ and $f(t)$, where $g(t) = \mathbf{L}^{-1}\{G(s)\}(t)$ for the transfer function $G(s)$ and where $f(t)$ is the forcing function, so that $\mathbf{L}^{-1}\{G(s)F(s)\}(t) = (g * f)(t) = x(t)$.

Lab 14: Tool Instructions

Laplace: Definition Tool

Parameter Sliders

Use the slider to change the value constant for s values between 0 and 4.

Press the mouse down on the slider knob for the parameter you want to change and drag the mouse back and forth, or click the mouse in the slider channel at the desired value for the parameter.

Laplace: Transformer Tool

Parameter Sliders

Use the slider to set the value constant for b between 0 and 4.

Press the mouse down on the slider knob and drag the mouse back and forth, or click the mouse in the slider channel at the desired value for the parameter.

Drawing Mode Buttons

Click the mouse on the **[L]** button to see the graphical output of the Laplace Transform of a selected function.

Click the mouse on the **[Inverse]** button to change the **[L]** button to the **[L^{-1}]** button

Click the mouse on the **[L^{-1}]** button to see the graphical output of the inverse Laplace Transform of a selected function.

Click the mouse on the **[Laplace]** button to change the **[L^{-1}]** button to the **[L]** button.

Other Buttons

Click the mouse on the **[Set Function]** button to choose a function.

Click on a function to select it.

Laplace: Derivative Tool

Drawing Mode Buttons

Click the mouse on the **[L]**, **[Differentiate]**, and/or **[Derivative Theorem]** buttons to see graphical output.

Other Buttons

Click the mouse on the **[Clear]** button to remove all output from the graphs.

Laplace: Vibrations and Poles Tool

Parameter Sliders

Use the slider to change the values for the parameter b.

Press the mouse down on the slider knob and drag the mouse back and forth, or click the mouse in the slider channel at the desired value for the parameter.

Drawing Mode Buttons

Click the mouse on the **[Start]** button to see graphical output.

Other Buttons

Click the **[Pause]** button to stop a trajectory without canceling it.

Click the **[Continue]** button to resume the motion of a paused trajectory.

Laplace: Translation Tool

Parameter Sliders

Use the slider to set the value constant for a values between –1 and 4.

Press the mouse down on the slider knob for the parameter you want to change and drag the mouse back and forth, or click the mouse in the slider channel at the desired value for the parameter.

Drawing Mode Buttons

Click the mouse on the **[L]** or the **[L⁻¹]** buttons to toggle between a function's graphical output and the graphical output of its Laplace Transform.

Laplace: Solver Tool

Drawing Mode Buttons

Click the mouse on the **[L]**, **[Solve]**, and **[L⁻¹]** arrow buttons to see algebraic and graphical output.

Other Buttons

Click the mouse on the function buttons below to choose a function, its initial conditions, and the type of damping.

Laplace: Shifting Theorem Tool

Parameter Sliders

Use the slider to set the value constant for a values between 1 and 4.

Press the mouse down on the slider knob for the parameter you want to change and drag the mouse back and forth, or click the mouse in the slider channel at the desired value for the parameter.

Drawing Mode Buttons

Click the mouse on the **[L]** button to see graphical output.

Laplace: Shift and Step Tool

Parameter Sliders

Use the slider to set the value constant for a values between 0 and 4.

Press the mouse down on the slider knob and drag the mouse back and forth, or click the mouse in the slider channel at the desired value for the parameter.

Laplace: Delta Function Tool

Parameter Sliders

Use the slider to set the value constants for a values between 0 and 3, and for h values between 0 and 4.

Press the mouse down on the slider knob for the parameter you want to change and drag the mouse back and forth, or click the mouse in the slider channel at the desired value for the parameter.

Laplace: Convolution Example Tool

Parameter Sliders

Use the slider to set the value constant for T values between 0 and 2.

Press the mouse down on the slider knob for the parameter you want to change and drag the mouse back and forth, or click the mouse in the slider channel at the desired value for the parameter.

Laplace: Convolution Theorem Tool

Drawing Mode Buttons

Click the mouse on the [L], [convolve], and/or [multiply] arrow buttons to see graphical output.

Other Buttons

Click the mouse on the [Clear] button to remove all output from the graphing plane.

Part

III   Linear Algebra

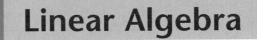

15 Linear Algebra

Tools Used in Lab 15
The Matrix Machine
The Eigen Engine

*A matrix **A** is not just a table of numbers. It is a machine that takes an input vector $\vec{v}$ and transforms it into an output vector $\mathbf{A}\vec{v}$ where $\mathbf{A}\vec{v}$ denotes multiplication of the vector by the matrix. Like an ad for a miracle diet, the "before" and "after" pictures of the vector are usually dramatically different! How does a matrix transform vectors?*

1. Visualizing Matrices as Transformations

The Matrix Machine tool allows you to see how an input vector (yellow) is stretched, rotated, inverted, projected, or otherwise transformed by a matrix to another vector (blue). Try this process for the following matrices.

1.1 Input the matrix $\mathbf{A} = \begin{bmatrix} 0 & 1 \\ -1 & 0 \end{bmatrix}$ by clicking the arrows to the left or right of each number to lower or raise the value. As you move the mouse around over the *xy*-plane, the mouse coordinates define the yellow input vector $\vec{v} = \begin{bmatrix} x \\ y \end{bmatrix}$. Then the tool plots the blue output vector $\vec{w} = \mathbf{A}\vec{v} = \begin{bmatrix} 0 & 1 \\ -1 & 0 \end{bmatrix}\begin{bmatrix} x \\ y \end{bmatrix} = \begin{bmatrix} y \\ -x \end{bmatrix}$. Watch how the output vector $\vec{w}$ changes.

How does the length of $\vec{w}$ compare with the length of $\vec{v}$?
Hint: Look at the circles.

How is the output vector related to the input vector? What is the geometrical effect of the transformation matrix **A**? Sketch some representative $\vec{v}$, $\mathbf{A}\vec{v}$ pairs.

1.2 Try $A = \begin{bmatrix} 1 & 2 \\ 0 & 1 \end{bmatrix}$. Sketch some representative $\bar{v}$, $A\bar{v}$ pairs. Drag the mouse around until you find an

input vector $\bar{v}$ that gets transformed into $\bar{w} = \begin{bmatrix} 0 \\ 1 \end{bmatrix}$. Find this $\bar{v}$ using algebra.

True or false? For any given $\bar{w}$, you can always find a $\bar{v}$ that gets transformed into that $\bar{w}$, and that $\bar{v}$ is unique (that is, there is only one $\bar{v}$ that will work). Explain your reasoning.

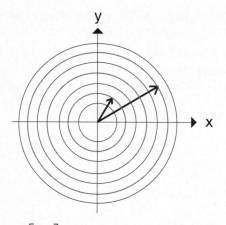

1.3 Try $A = \begin{bmatrix} 1 & 1 \\ 1 & 1 \end{bmatrix}$. Sketch some representative $\bar{v}$, $A\bar{v}$ pairs. By moving the mouse around and watching

the output, show that there are many different input vectors that all produce the output vector

$\bar{w} = \begin{bmatrix} 0 \\ 0 \end{bmatrix}$. (These input vectors are said to be in the nullspace, or kernel, of A.) Find an equation for

all the input vectors $\bar{v}$ in the nullspace.

Now find all the different input vectors that get transformed into $\bar{w} = \begin{bmatrix} 2 \\ 2 \end{bmatrix}$. How are these vectors related to the nullspace?

You must have noticed that the output vectors $\bar{v}$ always lie along a certain line. This set of all possible $\bar{v}$ is called the range of **A**. Find a formula for the vectors in the range of **A**.

2. Visualizing Eigenvectors and Eigenvalues

The Eigen Engine tool allows you to see what eigenvalues and eigenvectors mean geometrically. As you input a vector by dragging the mouse around, the tool first converts the vector into a unit vector $\bar{v}$, and then $\bar{v}$ is transformed by multiplying it by the matrix **A**. The resulting output vector $\bar{w} = \mathbf{A}\bar{v}$ is plotted— notice that it usually points in a different direction from $\bar{v}$.

However, for certain special choices of $\bar{v}$, it sometimes happens that $\bar{v}$ and $\bar{w}$ lie along the same line! When this happens, we say that $\bar{v}$ is an **eigenvector,** and the line on which $\bar{v}$ and $\bar{w}$ lie is the **eigen direction**. The screen freezes when you find an eigenvector and displays the **eigenvalue**, the factor by which the unit input vector was stretched or contracted. You can read off the magnitude of the eigenvalue as the radius on the polar grid. The sign of the eigenvalue is positive if $\bar{v}$ and $\bar{w}$ point in the same direction, it is negative if they point in opposite directions, and the eigenvalue is 0 if $\bar{w} = 0$. Click elsewhere on the graph to free up the vectors.

2.1   Find the eigenvectors and eigenvalues of the matrix $\mathbf{A} = \begin{bmatrix} 2 & 0 \\ 0 & 3 \end{bmatrix}$.

Show that if $\bar{v}$ is an eigenvector, so is $-\bar{v}$, and they have the same eigenvalue. Then prove that this is true in general for any matrix **A**.

2.2   Using **The Eigen Engine** tool, find the eigenvectors and eigenvalues of the following matrices. In some cases, you will find eigenvectors lying along two or more separate eigendirections, whereas in others you may find only one or possibly no eigendirections. Sketch the eigenvectors.

a. $\mathbf{A} = \begin{bmatrix} 1 & 2 \\ 0 & 1 \end{bmatrix}$

b. $\mathbf{A} = \begin{bmatrix} 1 & 1 \\ 1 & 1 \end{bmatrix}$

c. $\mathbf{A} = \begin{bmatrix} 0 & 1 \\ -1 & 0 \end{bmatrix}$

d. $\mathbf{A} = \begin{bmatrix} -2 & 0 \\ 0 & -2 \end{bmatrix}$

Lab 15: Tool Instructions

The Matrix Machine Tool

Input Vectors

Move the mouse over the xy plane to define the input vector and create the output vector.

Matrix Element Values

Click the pointers to the left and the right of the matrix elements to increase or decrease their values.

The Eigen Engine Tool

Input Vectors

Move the mouse over the xy plane to define the input vector and create the output vector. Both the input and the output vectors will freeze when they coincide with an eigenvector. Click the mouse away from the vectors to release them from the eigenvector.

Matrix Element Values

Click the arrows to the left and right of the matrix elements to increase and decrease their values respectively.

PARAMETER PLANE

THE BIFURCATION RAILROAD

GOOD FOR ONE FARE ONLY

LY 7, 1997

FRIDAY
0:00 AM

16   Linear Classification

Tools Used in Lab 16
Parameter Path Animation
Parameter Plane Input
Matrix Element Input
Four Animation Paths

We have four parameters in the planar system, $\dot{x} = ax + by$ and $\dot{y} = cx + dy$, but to get a parameter plane, we want only two! How do changes in location on the parameter plane affect the phase plane portraits? Take a tour and find out!

1. The Parameter Plane

The planar system of equations can be written

$$\begin{bmatrix} \dot{x} \\ \dot{y} \end{bmatrix} = \begin{bmatrix} a & b \\ c & d \end{bmatrix} \begin{bmatrix} x \\ y \end{bmatrix}, \quad (1)$$

or equivalently, $\dot{x} = \mathbf{A}x$, from which we obtain the second-order differential equation with constant coefficients called the trace, *tr*, and determinant, *det*.

$$\ddot{x} - \underbrace{(a+d)}_{tr\mathbf{A}}\dot{x} + \underbrace{(ad-bc)}_{det\mathbf{A}}x = 0 \quad (2)$$

with the characteristic equation:

$$\lambda^2 - tr\mathbf{A}\lambda + det\mathbf{A} = 0 \quad (3)$$

We can get the roots of this equation via the quadratic formula:

$$\lambda = \frac{tr\mathbf{A} \pm \sqrt{(tr\mathbf{A})^2 - 4det\mathbf{A}}}{2} \quad (4)$$

so that the sign of the discriminant, $\Delta = (tr\mathbf{A})^2 - 4det\mathbf{A}$, determines whether there are two real unequal roots $\lambda_1 \neq \lambda_2$, one repeated real root λ_1, or no real roots.

1.1 True or false? Two different matrices cannot have the same trace, determinant, and characteristic equation. Explain if true, or give a counterexample if false.

We construct a **parameter plane** with $tr\mathbf{A}$ for the horizontal axis and $det\mathbf{A}$ for the vertical axis. The parabola $det\mathbf{A} = (tr\mathbf{A})^2/4$ is the locus where the discriminant is 0.

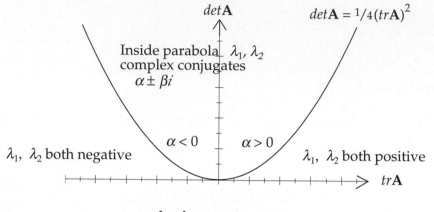

2. Take the Tour

Start the **Parameter Path Animation** tool and begin the tour. Think of riding a railroad car along a path in the parameter plane and watching the phase plane scenery go by. Whenever a small change in a parameter produces a marked change in the behavior of the trajectories, we say a bifurcation has occurred. As you can see here, the bifurcations occur as we move from one colored zone to the next. Several important concepts are illustrated here. We will examine them more carefully using other tools.

What is a fixed point? The motion of a point along a trajectory in the phase plane can be described by means of $(\dot{x}, \dot{y})$, which tells us how the x-coordinate and y-coordinate are changing with respect to time. Imagine a point flowing along a trajectory. At any point (x, y) where $(\dot{x}, \dot{y}) = (0, 0)$, we have a fixed point—there is no change. Our imaginary point would also remain at the origin if placed there. You can see that the system is in equilibrium at a fixed point. The question becomes "What kind of equilibrium?"

2.1 Every linear system of the form (1) will have a fixed point at the origin. Can such a system have more than one fixed point? Explain.

The behaviors of the trajectories in the neighborhood of the fixed point allow us to classify the fixed point. Is it an **attracting** or **repelling** fixed point? In other words, do trajectories in the neighborhood of the fixed point approach it or go away from it as $t \to \infty$? If neither, it is called a **neutral** fixed point. There are many ways of describing fixed points, but to get a better feel for their meanings, do the following exercise.

2.2 Use the **Parameter Plane Input** tool. The origin is always a fixed point. Select a pair of parameters by clicking in one of the colored zones. Look at the resulting values of the λ's, and make a phase portrait by clicking on the initial values for a few trajectories in the phase plane. Make a rough sketch of the portraits below the values for the λ's. Note that each phase portrait is named on the screen. Include those names as labels on your sketches. Label the equilibrium points as attracting, neutral, or repelling. Include only the major colored zones and the positive vertical axis. We will look at the other borderline cases later.

a. $\lambda_1 > \lambda_2 > 0$

b. $\lambda_1 < \lambda_2 < 0$

c. $\lambda_1 < 0, \quad \lambda_2 > 0,$
    (or vice versa)

d. $\lambda = \pm \beta i$

   $\beta \neq 0$

e. $\lambda = \alpha \pm \beta i$
    $\alpha > 0, \beta \neq 0$

f. $\lambda = \alpha \pm \beta i$
    $\alpha < 0, \beta \neq 0$

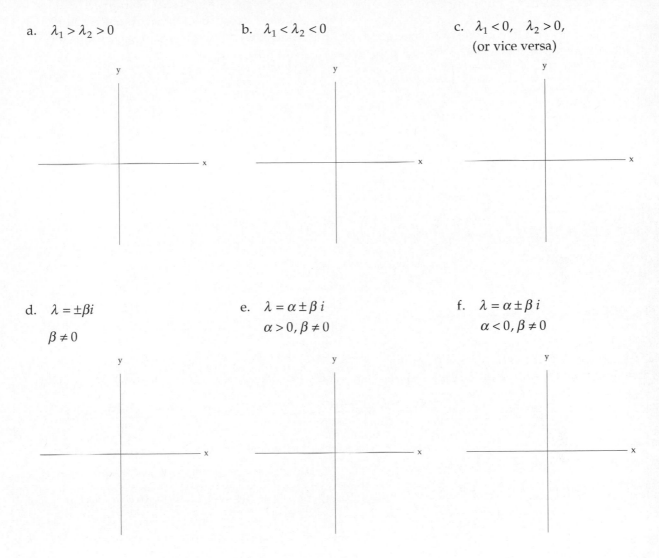

g. Which of these phase portraits has "closed orbits"? This behavior characterizes phase portraits for periodic motion.

h. Which one of these phase portraits represents simple harmonic oscillations?

Transitions

2.3   Use the **Parameter Plane Input** tool to observe the changes in the pictures and the roots, λ, of the characteristic Equation (3) that occur as you move the mouse across boundaries from one colored zone to the next. Set parameter plane points on either side of each boundary and set some points in the xy-plane to observe trajectories.

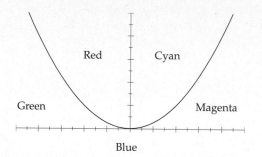

From blue to magenta:

From magenta to cyan (lighter blue-green):

From cyan to white (positive vertical axis):

From white to red:

From red to green:

From green to blue:

3. A Real Example

3.1   The Mass-Spring Problem, from Labs 9 and 10

a.   Consider the mass-spring problem with damping, where the mass $m = 1$, the damping constant is δ, and the spring constant $k = 1$. The resulting second-order differential equation is $\ddot{x} + \delta\dot{x} + x = 0$. Write this equation as a planar system of linear differential equations in the standard fashion:

$$\dot{x} = y$$

$$\dot{y} =$$

b.   What is the corresponding matrix **A** so that $\dot{x} = \mathbf{A}x$?

c.   Use the **Parameter Plane Input** tool to check the phase portraits for the fixed point as you change the values of matrix element δ from –3 to 3. What is the trace of matrix **A** in terms of δ?

d.   What is the color of the zone or border that corresponds to an underdamped oscillator? Sketch the phase portrait.

e.   What is the color of the zone or border that corresponds to critical damping? Sketch the phase portrait.

f.   What is the color of the zone or border that corresponds to the overdamped case? Sketch the phase portrait.

g.   Can you interpret the results for negative values of element δ? (*Hint:* When $\delta < 0, \ tr\mathbf{A} > 0$.)

h.   Suppose there is no damping. What kind of fixed point would the origin be?

The matrices used so far have had $\begin{bmatrix} 0 & 1 \end{bmatrix}$ in the upper row. This choice reflects the fact that a linear second-order differential equation can always be rewritten so that $\dot{x} = y$ is the first equation in the system. The **Matrix Element Input** tool allows you to choose other real matrix entries. We will use this feature to investigate a few more cases.

4. Borderline Cases

The most important borderline case divides regions of stable and unstable spirals, sometimes called **spiral sinks** and **spiral sources**, respectively. In the phase plane the corresponding trajectories are ellipses. Recall that **closed orbits** are indicative of periodic motion. The fixed point for this case is a called a **neutral center**; neutral because the trajectories are neither attracted to nor repelled by the fixed point. This situation characterizes frictionless, or **conservative**, systems.

4.1   Use the **Parameter Plane Input** tool to examine the change from spiral to circle to spiral again as you move the cursor across the positive vertical axis. If your touch is exceedingly fine you can set a parameter point that fills the xy graph with a slowly decreasing spiral as the orbit just misses being periodic. Note that the trajectory stops when a spiral takes too much time. You can also interrupt a prolonged trajectory with a mouse click.

4.2   Use the tool called **Four Animation Paths** to observe the behavior of the other borderline cases. Sketch phase portraits of typical behaviors for each of the cases along the parabola. Label them with the names on the screen (for example, degenerate node).

a. $\lambda_1 = \lambda_2 = \lambda > 0$ b. $\lambda_1 = \lambda_2 = \lambda = 0$ c. $\lambda_1 = \lambda_2 = \lambda < 0$

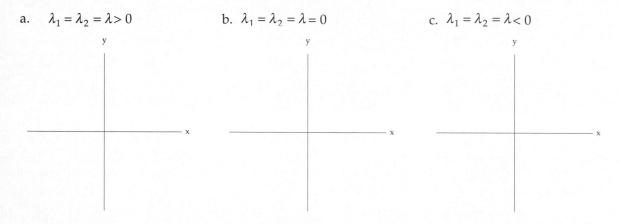

Note that in **(a)** and **(c)** another portrait exists. We need to examine this case more closely. If you have done Lab 15, Linear Algebra, you know this part already. If not, we introduce the basic idea here, in section 5. For more information, refer to Lab 15.

Observe that moving along the horizontal axis, where $det\,\mathbf{A} = 0$, gives us a line or a plane of **non-isolated fixed points**. You can think of a **degenerate node** as being the borderline case between a spiral and a node, where one tries to deform into the other.

5. The Role of Eigenvectors in Borderline Cases

An **eigenvector** for a matrix **A** is a nonzero vector $\bar{v}$ for which

$$\mathbf{A}\bar{v} = \lambda\bar{v}$$

where the **eigenvalue,** λ, is a constant. An eigenvector may be stretched, shrunk, reversed, or left alone by the matrix, but it is never rotated. A given eigenvalue has many **eigenvectors.** The eigenvectors for a given λ form a vector space called, not too surprisingly, the **eigenspace.**

Now look at what happens in the two cases with repeated (nonzero) eigenvalues, $\lambda_1 = \lambda_2 = \lambda \neq 0$.

Case 1. There is only one linearly independent eigenvector. We've seen this case. It is by far the most common. The phase portrait shows that the fixed point is a degenerate node.

Case 2. There are two linearly independent eigenvectors, and since there is only one two-dimensional vector space containing these eigenvectors, every vector must be an eigenvector. That means that every vector gets stretched or shrunk, etc., in the same way.

5.1   Use the **Matrix Element Input** tool to try the matrix $\mathbf{A} = \begin{bmatrix} \lambda & 0 \\ 0 & \lambda \end{bmatrix}$ for any nonzero λ.

   a.   Sketch the phase plane portrait. This is called a **star node**. Is it attracting or repelling (that is, is λ positive or negative)? How is this fact related to the sign of λ?

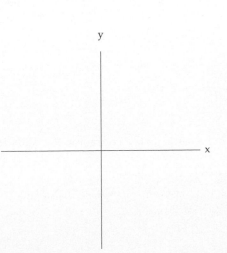

   b.   Now, using the same λ, try the matrix $\mathbf{A} = \begin{bmatrix} \lambda & k \\ 0 & \lambda \end{bmatrix}$ and vary k away from zero by increments until you get a nice degenerate node. Write your final matrix below, and sketch the phase portrait.

6. Additional Exercises

6.1   Using the **Matrix Element Input** tool to classify the fixed point (or points), find the roots, λ, of the characteristic equation, and sketch the phase portraits for the various matrices **A** where $\dot{\bar{x}} = \mathbf{A}\bar{x}$. (Note that these are the same problems as those in the last part of Lab 15, Linear Algebra.)

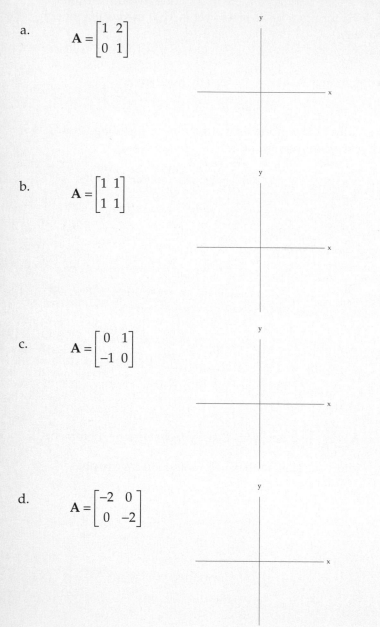

a. $\mathbf{A} = \begin{bmatrix} 1 & 2 \\ 0 & 1 \end{bmatrix}$

b. $\mathbf{A} = \begin{bmatrix} 1 & 1 \\ 1 & 1 \end{bmatrix}$

c. $\mathbf{A} = \begin{bmatrix} 0 & 1 \\ -1 & 0 \end{bmatrix}$

d. $\mathbf{A} = \begin{bmatrix} -2 & 0 \\ 0 & -2 \end{bmatrix}$

6.2   Show that the characteristic Equation (3) is exactly the same as the characteristic equation obtained from

$$det\begin{bmatrix} a - \lambda & b \\ c & d - \lambda \end{bmatrix} = 0$$

Lab 16: Tool Instructions

Parameter Path Animation Tool

Buttons
Click on the arrow buttons to control the animation sequence according to the parameter path. Use the double arrow buttons to play the sequence forward and backward. Use the single arrow buttons to advance and reverse the sequence one frame at a time. To stop a play sequence, use a single arrow button.

Parameter Plane Input Tool

Parameter Plane
Move the mouse over the left parameter plane to change the trace and determinant.
Click on the parameter plane to set the determinant and the trace and define the matrix, then click on the xy plane to the right to start trajectories. Click on the parameter plane again to release the parameter point and select a new trace and determinant.

Setting Initial Conditions
After setting a point on the parameter plane, click the mouse on the xy-graphing plane to set the initial conditions for a trajectory.
Clicking in the xy-plane while the trajectory is being drawn will start a new trajectory.

Buttons
Click the mouse on the [Clear] button to remove all the trajectories from the xy-graph.

Matrix Element Input Tool

Matrix Element Values
Click on a matrix element button in the lower-left corner of the screen to activate the text editor. Type in new values using the right and left arrow keys, the [Delete] key, and the number keys. Press return or click elsewhere to exit the text editor and set the matrix.

Setting Initial Conditions
Click the mouse on the graphing plane to set the initial conditions for a trajectory.
Clicking while a trajectory is being drawn will stop the trajectory.

Buttons
Click the mouse on the [Clear] button to remove all trajectories from the xy-graph.
Click the mouse on the [Draw Field] button to draw a grid of vectors over the xy-graph.

Four Animation Paths Tool

Buttons
Click on the arrow buttons to control the animation sequence of the parameter path. Use the double arrow buttons to play the sequence forward and backward. Use the single arrow buttons to advance and reverse the sequence one frame at a time. To stop a sequence, use the single arrow button.
Click on the path buttons in the lower-left corner of the screen to choose the path on the parameter plane that defines the animation sequence—parabolic, horizontal, vertical, or rectangular.

Part IV

Systems of Differential Equations

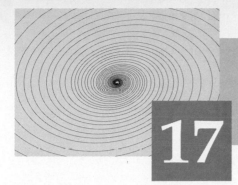

Graphing Two-Dimensional Equations

17

What do second-order equations have in common with systems of two first-order differential equations? Why are phase planes and vector fields so important? How do they relate to x(t) and y(t) time series?

Second-order differential equations and systems of two first-order equations are both problems in two dimensions. If a solution can be written as an equation, you need a two-parameter family of equations to describe all the solutions. Each individual solution must be specified by a two-dimensional condition, usually an initial condition.

For a second-order equation such as $mx'' + bx' + kx = f(t$, an initial condition would describe both $x(0)$ and $x'(0)$.

For a system of two first-order equations such as $dx/dt = f(x,y)$, $dy/dt = g(x,y)$, an initial condition would describe both $x(0)$ and $y(0)$.

Any second-order equation can be written as a system by assigning $x' = y$ and then noting that $y' = x''$. Hence for our sample equation above, $y' = f/m - by/m - kx/m$. This procedure gives x' and y' each as functions of x and y.

Many systems can similarly be transformed to a second-order equation by solving for one variable in terms of the others and substituting; for example, if you can solve the first equation for y in terms of x and x', then you can substitute in the second for y and y', giving a resulting equation in x'', x', and x.

You can deal with these manipulations outside of the lab. Our purpose here is just to stress that these two problems are essentially the same, so that we can explore in a single discussion the various graphs that result, and how they relate.

1. The Graphs

1.1   The most famous second-order equation is also one of the simplest: $x'' = -x$, describing simple harmonic motion. You can review it with the **Simple Harmonic Oscillator** tool in Lab 9, Linear Oscillators: Free Response.

a. What are the analytic solutions to $x'' = -x$?

b. How do you write this equation as a system of first-order equations?

Note: From this point on, we will use the y label, with the understanding that in the case of a second-order equation, you can replace x' by y.

1.2   As you have seen in the labs of Part 2 for second-order equations, we can no longer just draw solutions $x(t)$ from a slope field, as we did for first-order equations. One way to write the analytic solution is

$$x(t) = A \cos (t + B). \tag{1}$$

a. Confirm algebraically that this is equivalent to your answer to Exercise **1.1a**.

b. If you want to graph more than one $x(t)$ solution at a time, the tx graph quickly gets complicated. Confirm that this tx graph of just a few of these solutions indeed matches Equation (1), by estimating A and B values for the numbered solution curves.

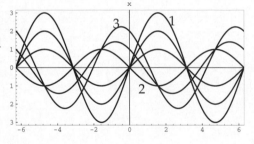

(1) $A_1 \approx$ $B_1 \approx$
(2) $A_2 \approx$ $B_2 \approx$
(3) $A_3 \approx$ $B_3 \approx$

What is the common period of all the curves on the tx graph?

c. Differentiate Equation (1) to get a formula for $y(t) = x'(t)$ in this example.

d. Confirm that this ty graph comes from the tx graph by matching a couple of curves. That is, use your estimated (A,B) values to state the corresponding initial conditions, and identify each of the three associated x' curves on the second graph.

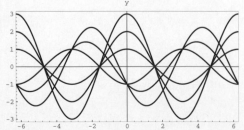

(1) $x(0) =$ $y(0) =$
(2) $x(0) =$ $y(0) =$
(3) $x(0) =$ $y(0) =$

Alternatively, you can accomplish the matching by using the fact that the $y(t)$ curves give the slopes of the $x(t)$ curves, at each value of t.

What period(s) are represented by the $y(t)$ curves?

1.3 Observe that on the tx and ty graphs, just a few solutions make quite a mess. But if you put the tx and ty information together as parametric equations, you can get a useful xy graph, called the **phase portrait**. Open the **Parametric to Cartesian** tool for a careful demonstration of how this is done for a single solution.

This graph, the xy graph associated with the tx and ty graphs in Exercise **1.2**, is much cleaner!

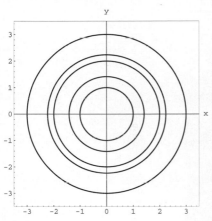

a. Add arrowheads to show the proper direction of the trajectories, which can be determined directly from the differential equations.

b. You may note, however, that one less curve appears on this phase portrait than in the tx and ty graphs illustrated. Which two tx curves give the same xy curve? Mark them with a colored pencil in Exercise **1.2** and do the same for the corresponding ty curves.

c. Consider the advantages and disadvantages of each of the three graphs presented. Each is most useful for different purposes. List at least two advantages and one disadvantage for each case.

Pros and cons of xy:

Pros and cons of tx and ty:

2. Making the Phase Plane

Note that the independent variable t does not show explicitly on the phase plane. Experiment with the **Phase Plane Drawing** tool to help answer the question "Where is t?" This tool allows you to draw a curve on the xy-plane using the mouse; as you draw, the corresponding $x(t)$ and $y(t)$ graphs appear.

2.1 a. Why is the tx-plane at the lower left rotated from its normal position?

b. Try drawing a semicircle about the origin in the upper half of the xy-plane, and observe the $x(t)$ and $y(t)$ graphs that result. Are they what you expected?

c. What do you think will happen on these graphs if you draw the semicircle in the opposite direction?

d. If you draw it twice as fast?

e. Try drawing a spiral from a boundary point to the origin. What sort of physical system might this represent?

2.2   Make, sketch, and explain an experiment of your own using this tool:

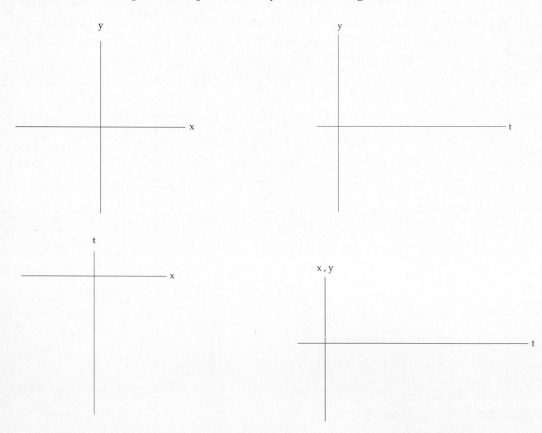

This **Phase Plane Drawing** tool is really just a Cartesian-to-parametric illustration—there is no differential equation involved, but it gives a good feeling for the connection between the graphs. In the **Parametric to Cartesian** tool, the phase point is visually linked to the time series coordinates for x and y. A system of differential equations can be solved numerically to get $x(t)$ and $y(t)$, then the xy phase portrait is constructed directly from these.

3. Vector Fields for Autonomous Systems

Having established that phase planes are what we often want to look at, we take a moment to observe their construction directly from a system of two differential equations, rather than from the functions $x(t)$ and $y(t)$. It is similar to the construction of a slope field for a one-dimensional differential equation, but there are some differences, such as:

- the vectors may point in any direction, not just left to right as in a slope field
- the curves that flow through a vector field according to the differential equation are called **trajectories,** distinct from the *time series* graphed in the *tx*- and *ty*-planes.

3.1 Use the **Vector Fields** tool to experiment with several different systems. Set enough vectors on the phase plane to visualize the flow of trajectories. Make a quick sketch of a few trajectories for each of the two systems below.

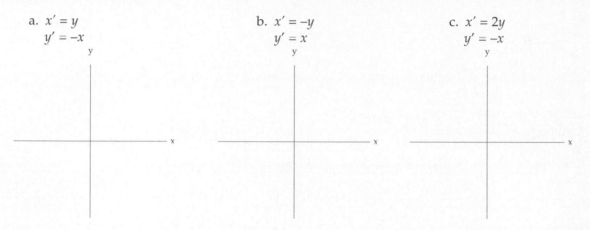

a. $x' = y$
   $y' = -x$

b. $x' = -y$
   $y' = x$

c. $x' = 2y$
   $y' = -x$

Compare graphs and equations for a, b, and c. How do the differences in the equations explain the differences in the graphs?

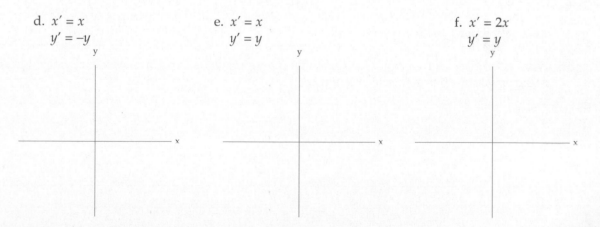

d. $x' = x$
   $y' = -y$

e. $x' = x$
   $y' = y$

f. $x' = 2x$
   $y' = y$

Compare graphs and equations for d, e, and f. How do the differences in the equations explain the differences in the graphs? Look carefully at the ways in which your answers are similar to or different from those in a, b, and c.

g. $x' = 0$
   $y' = y$

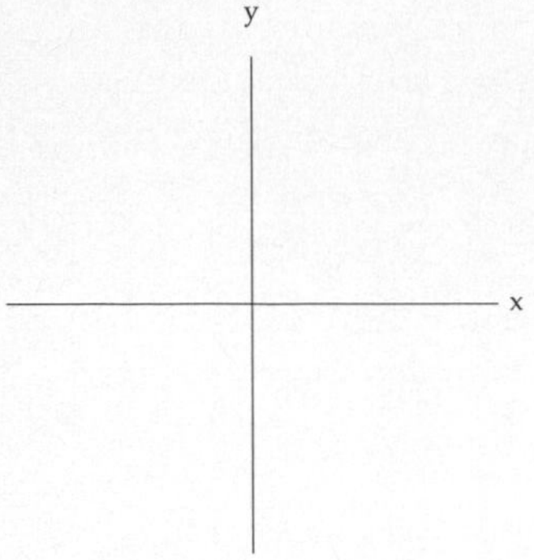

h. $x' = y$
   $y' = y - x$

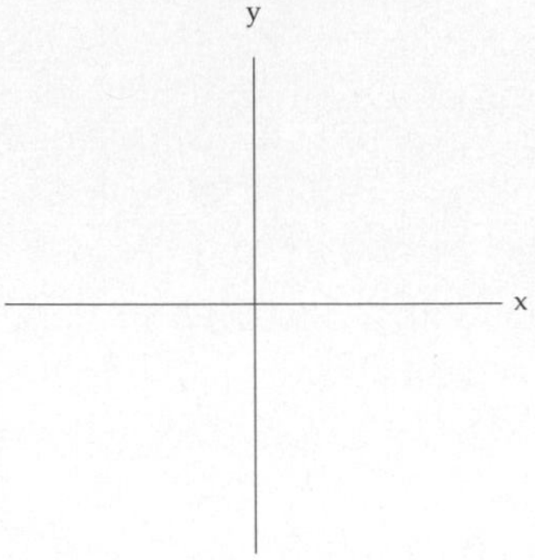

i. $x' = x - xy$
   $y' = xy/2 - y$

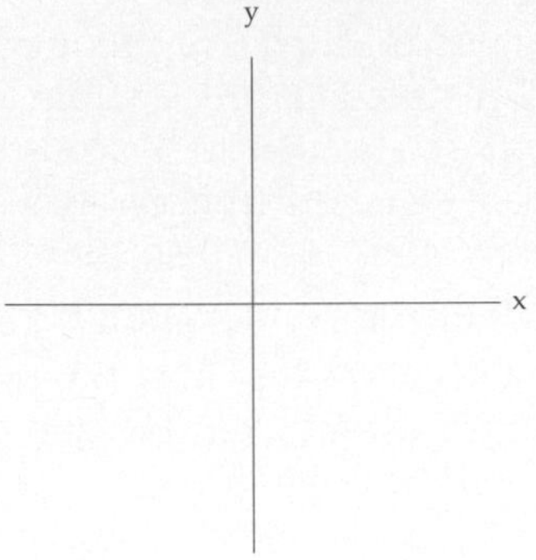

What does an equilibrium look like on a phase plane?

What do the equations look like at an equilibrium?

How many equilibria are there in each of the above graphs, a–i?

3.2   When the time step is 1, the vectors with components dx/dt and dy/dt are drawn to the correct scale, and their magnitude tells the *speed* of the trajectory.

     a.   It is easy to see that the speed is slowest at the equilibria or fixed points, where $x' = y' = 0$. For the nine equations in Exercise **3.1**, try to find where the speed is maximized for each system and mark the graphs (with a different color).

     b.   Write a sentence or two explaining why the magnitude of vectors in most vector fields is drawn with constant rather than scaled lengths.

Note: In many software programs, vector fields are drawn without arrowheads. This may not be a great sin if there are arrows drawn on the trajectories, but it is important to realize that in a two-dimensional system, at each point there really is a *vector* with both a magnitude and a direction, as determined by the differential equation.

4. Relating the Graphs

4.1   For each of the equations from Exercise **3.1**, sketch a typical phase plane trajectory from your graph in Exercise **3.1**, and mark a starting point. Then, without the computer, sketch what you would expect the *tx* and *ty* curves to look like for that trajectory. Use the **Two-Dimensional Equations** tool to see if you were right. If you thought of some aspects, but others eluded you, correct your trajectory in a different color and add an explanation. (It is expected that this will happen—it's the explanation of the correction that is important. Show what happened, without erasing.)

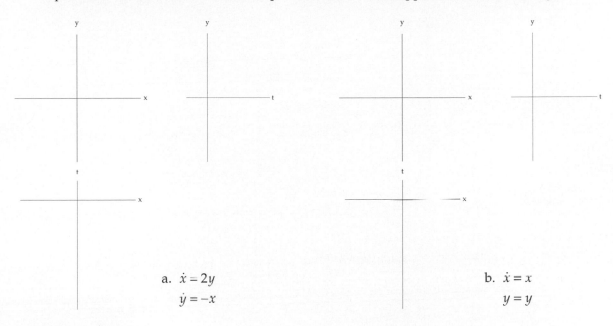

a. $\dot{x} = 2y$
   $\dot{y} = -x$

b. $\dot{x} = x$
   $\dot{y} = y$

Explanations of disparities: Explanations of disparities:

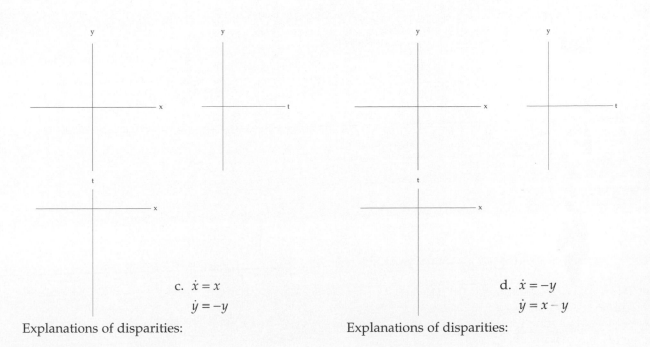

c. $\dot{x} = x$
   $\dot{y} = -y$

d. $\dot{x} = -y$
   $\dot{y} = x - y$

Explanations of disparities: Explanations of disparities:

4.2   Now that you have gained a bit of insight into what to expect for the *tx* and *ty* graphs, look at the following graphs that were drawn simultaneously, and *without the computer*, try to label the *tx* and *ty* curves that match each of the labeled *xy* trajectories.

(*Note:* In some cases the *tx* and *ty* curves may have gone a bit longer than shows on the *xy*-plane.) If you get stuck, you could turn to the **Two-Dimensional Equations** tool for help on some of these, but you should still try the next one by hand first. The goal is to become good at this kind of matching, because it comes in very handy.

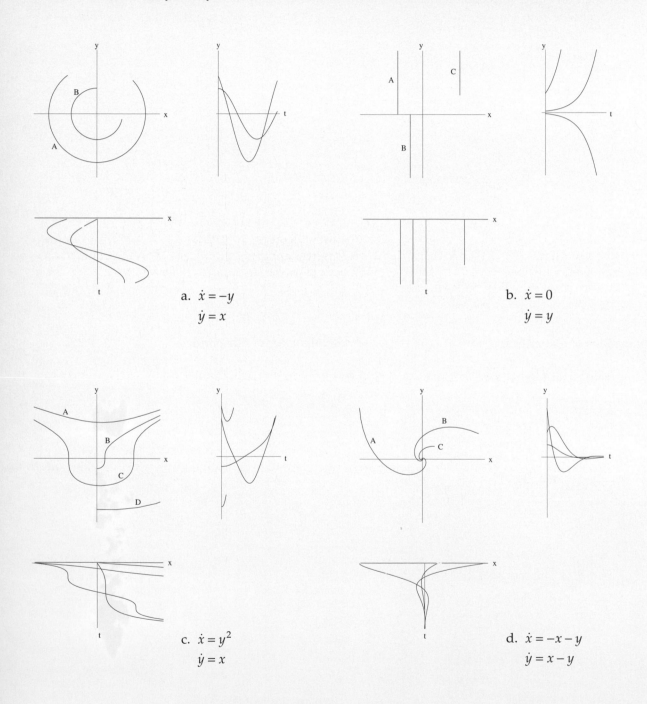

a. $\dot{x} = -y$
   $\dot{y} = x$

b. $\dot{x} = 0$
   $\dot{y} = y$

c. $\dot{x} = y^2$
   $\dot{y} = x$

d. $\dot{x} = -x - y$
   $\dot{y} = x - y$

5. Nullclines

The isoclines used to draw slope fields are useful also with vector fields and phase planes, but now that more variables are involved, it is usually easiest just to draw the **nullclines**:

- the isocline of horizontal slopes is where $dy/dt = 0$;
- the isocline of vertical slopes is where $dx/dt = 0$.

5.1   Using a new color, locate the nullclines on each of the phase plane portraits drawn in Exercise **3.1**.

5.2   Equation i in Exercise **3.7** represents an example of a predator-prey system, such as appears in the **Lotka-Volterra** tool from Lab 21, Predator Prey Population Models, with $x = H$ and $y = P$. Open that tool, which focuses on more of the first quadrant. Move the nullclines slightly by moving the sliders. What happens to the overall field and to the equilibria?

Imagine walking across the vector field from left to right. What happens to the vectors as you cross the isocline of horizontal slopes?

What changes occur in the vectors if you now walk across the vector field from top to bottom, when you cross the isocline of vertical slopes?

While you are in this example, watch a trajectory draw and comment on its speed. Where does it go fast? Where does it go slow? Why?

The regions separated by the nullclines can be qualitatively analyzed (by the differential equations) to tell whether the vectors point right or left, up or down. Equilibria, designated by circles, will occur where an isocline of horizontal slopes meets an isocline of vertical slopes. Here a typical nullcline sketch is given with a phase portrait for a similar example.

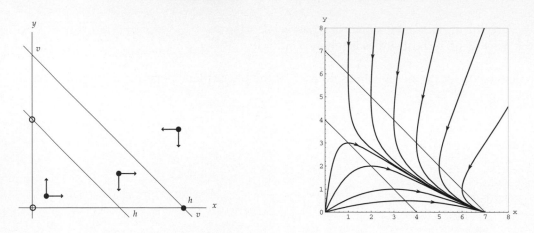

5.3　On the phase portrait above, verify that the trajectories pass through the nullclines with proper slopes and confirm that the directions of the trajectories indeed fall in the quadrants predicted on the left-hand nullcline graph.

Note: Other examples of this sort can be found in Lab 22, Competing Species Population Models, in the **Competitive Exclusion** tool, where $x = N_1$, $y = N_2$, and (as in the **Lotka-Volterra** tool) the nullclines and vector field can be changed by moving sliders for the various coefficients.

5.4　Naturally, nullclines are not always straight lines. The following equations include excellent examples. With a different color, sketch the nullclines on the following phase portraits, and mark the general left-right and up-down directions in the various regions. Then check your results with either the **Vector Fields** tool or the **Two Dimensional Equations** tool. The last example is a special case—explain what happens.

a. $x' = y^2$　　　　　　b. $x' = x - y$　　　　　　c. $x' = -x + y$
　　$y' = x$　　　　　　　　$y' = y - x^2$　　　　　　　　$y' = x - y$

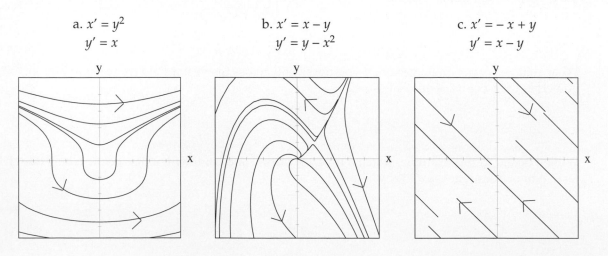

Lab 17: Tool Instructions

Simple Harmonic Oscillator Tool

Setting Initial Conditions

Click the mouse on the phase plane to set the initial position and the initial velocity, or click the mouse on the time series graph to set the initial position (initial velocity defaults to zero). Clicking while a trajectory is being drawn will stop the trajectory.

Time Series Buttons

The buttons labeled
[ ] **position**
[ ] **velocity**
[ ] **acceleration**
toggle the time series graphs on and off.

Other Buttons

Click the [**Pause**] button to stop a trajectory without canceling it.
Click the [**Continue**] button to resume the motion of a paused trajectory.
Click the mouse on the [**Clear**] button to remove all output from the graphs.

Parametric to Cartesian Tool

Buttons

Click the mouse on the [**Project x**] button to show the corresponding position of x on the phase plane.
Click the mouse on the [**Project y**] button to show the corresponding position of y on the phase plane.
Click the mouse on the [**Project x and y**] button to show the corresponding positions of both x and y on the phase plane as phase point coordinates.
Click the [**Pause**] button to stop a trajectory without canceling it.
Click the [**Continue**] button to resume the motion of a paused trajectory.

Phase Plane Drawing Tool

Drawing in the Phase Plane

Press and hold down the mouse button, then move the mouse to draw in the xy phase plane.

Buttons

Click the mouse on the [**Clear**] button to remove all output from the graphs.

Vector Fields Tool

Setting Initial Conditions

Click the mouse on the xy graphing plane to set the initial conditions for a trajectory or a point for a vector.
Clicking while a trajectory is being drawn will stop the trajectory.

Equations

Click the arrow button to the left of the equations to pop up the list of equations.
Click an equation to select it.

Drawing Mode Buttons

Click the mouse on the **[Vectors]** button to set vectors when you click on the xy plane.
Click the mouse on the **[Solutions]** to display a solution curve when you click on the xy plane.

Time Step Buttons

Click the mouse on a button in the Δt list to set the time step for vectors and trajectories.

Other Buttons

Click the mouse on the **[Clear]** button to remove all output from the graph planes.
Click the mouse on the **[Draw Field]** button to draw a grid of vectors over the xy graphs.

Two Dimensional Equations Tool

Setting initial conditions

Click the mouse on any of the three graphing planes to set the initial conditions for a trajectory.
Clicking while a trajectory is being drawn will stop the trajectory.

Equations

Click the arrow button to the left of the equations to pop up the list of equations.
Click an equation to select it.

Buttons

Click the mouse on the **[Clear]** button to remove all output from the graphs.
Click the mouse on the **[Draw Field]** button to draw a grid of vectors over the xy graphs.
Click the mouse on the **[Pause]** button to stop a trajectory without canceling it.
Click the mouse on the **[Continue]** button to resume the motion of the paused trajectory.

Lotka-Volterra Tool

Setting Initial Conditions

The initial conditions for the active trajectory are displayed to the right of the HP graphing plane.
They are set using either the mouse or the keyboard.

1. Click the mouse on the HP graphing plane to set $H(0)$ and $P(0)$.
2. Click the mouse on the value for one of the initial conditions displayed beside the graph to activate a keyboard editor. Set a new number using the number keys, the right and left arrow keys, and the Delete key. Press the **[Return]** key or click the mouse away from the number to leave the editor, set the initial value, and start the trajectory.
3. Clicking while a trajectory is being drawn will stop the trajectory.

Parameter Sliders

Use the sliders to set the growth rate a, the predation rate b, the predator mortality rate c, and the food conversion rate d.
Press the mouse down on the slider knob for the parameter you want to change and drag the mouse back and forth, or click the mouse in the slider channel at the desired value for the parameter.

Buttons

Click the mouse on the **[Clear]** button to remove all output from the graphs.
Click the mouse on the **[Pause]** button to stop a trajectory without canceling it.
Click the mouse on the **[Continue]** button to resume the motion of the paused trajectory.

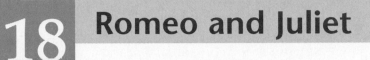

18 Romeo and Juliet

Every love affair has its ups and downs over time . . . so can it be modeled by differential equations?

Disclaimer: Only the names are the same.

1. The General Case

The situation is fictional but perhaps familiar. The gender of the participants can be changed to fit personal situations. The problem was invented by Steven Strogatz [SS1], the "treatment program" instituted by Bjørn Felsager [MF], and further alterations and inventions can be made by you after you understand the nature of the problem. You will see that love is sometimes best modeled by an oscillator . . . perhaps, alas, a damped oscillator.

Although love by its nature may be nonlinear, we restrict our attention to the linear case. This leaves the door open for further research.

Let x and y be functions of time, where x denotes Romeo's love for Juliet and y denotes Juliet's love for Romeo.

$$\frac{dx}{dt} = ax + by$$
$$\frac{dy}{dt} = cx + dy$$

where a, b, c, and d are constants.

Note that these equations can be written in matrix form as

$$\begin{bmatrix} \dot{x} \\ \dot{y} \end{bmatrix} = \begin{bmatrix} a & b \\ c & d \end{bmatrix} \begin{bmatrix} x \\ y \end{bmatrix}$$

Setting the Stage

Romeo's love for Juliet cools in proportion to her love for him. Juliet's love for Romeo grows in proportion to his love for her. Time is measured in days (0–50) and their love is measured on a scale from –5 to 5, where 0 is indifference.

Hysterical hatred	Disgust	Indifference	Sweet attraction	Ecstatic love
–5	–2.5	0	2.5	5

When first they meet, Romeo is immediately attracted to Juliet, but she is as yet indifferent. Soon, however, the tide will turn . . .

The **Romeo and Juliet** tool allows you to change elements in the matrix by clicking on the arrows beside the values to make them larger or smaller. You can set the initial conditions by clicking on an appropriate spot in the phase plane. The initial conditions described above are $x(0) = 2$, $y(0) = 0$. Note that the form of the equations given in the tool uses $(x + h)$ and $(y + k)$ instead of x and y. This is useful for Exercise **3.1**.

1.1 Set the matrix entries in the **Romeo and Juliet** tool to correspond to the system of equations below. Use initial conditions $x(0) = 2$, $y(0) = 0$.

$$\frac{dx}{dt} = -0.2y$$

$$\frac{dy}{dt} = 0.8x$$

a. Look at the phase plane and time series for x and y to analyze the ensuing relationship. How would you characterize their relationship?

b. Look at the phase plane. What percentage of the time do they experience mutually positive feelings for each other?

c. Examine the time series for x and y. What are Juliet's feelings for Romeo when he is most attracted to her?

   What are Romeo's feelings for Juliet when she is most attracted to him?

d. What is the maximum attraction or love that Romeo feels for Juliet? What is the maximum attraction or love that she feels for him? Give your answers in both numbers and words.

e. What is the period of their emotional cycles? Do they have the same period?

f. What is the time between Romeo's and Juliet's emotional peaks (that is, the shortest time)?

2. The story continues . . .

Juliet's ride on an emotional roller coaster begins to disorient her. She is distracted, losing sleep and forgetting to do her homework, so she goes to a counselor for advice. The counselor decides that she is over-responding to the emotional stimuli and puts her on a tranquilizer. The new equations follow.

$$\frac{dx}{dt} = -0.2y$$

$$\frac{dy}{dt} = 0.8x - 0.1y$$

2.1   Adjust the matrix entries to reflect the new situation. Use the same initial conditions as in Exercise **1.1**: $(x(0) = 2, y(0) = 0)$.

a.   Sketch the phase plane and the time series. What happens? Can we say that the tranquilizer has a damping effect on the relationship?

b.   Suppose the counselor now tries instead to increase the responsiveness of Romeo to Juliet by giving him a stimulant. What do you expect to happen to the relationship? Using the equations below and the same initial conditions, does the graph confirm your prediction?

$$\frac{dx}{dt} = -0.2y + 0.1x$$

$$\frac{dy}{dt} = 0.8x$$

3. And the story continues . . .

Romeo and Juliet are appalled at the changes in their relationship and immediately change counselors. The new counselor, wiser and more understanding, realizes that a shift in attitudes must take place. They stop taking pills and start to learn new patterns of response to each other. Romeo learns to accept Juliet's love and now his love begins to decrease only if she becomes overly affectionate ($y > 2$). Juliet learns to control her responses and her love grows only when Romeo becomes very affectionate ($x > 2$). The equations become

$$\frac{dx}{dt} = -0.2(y - 2)$$

$$\frac{dy}{dt} = 0.8(x - 2)$$

3.1   Change the matrix entries accordingly and the column vector $\begin{bmatrix} x \\ y \end{bmatrix}$ to $\begin{bmatrix} x - 2 \\ y - 2 \end{bmatrix}$. Use the same initial conditions. Sketch the phase portrait and time series. Describe what happens. Have they found happiness at last?

4. Variations on the Theme

Of course, there are many variations on the modeling of love affairs with differential equations. We can include time dependence, such as the onset of spring, or triangulate with a new love or an old flame.

The **Romeo and Juliet** tool is designed to handle only linear equations, but within these limitations there are many variations. Some of these variations are included in the exercises that follow. For further variations, consult the references listed at the end of the lab.

4.1   One of the references [SS2, 138-144] examines the effects of the signs of the parameters on the course of the love affairs. Pick one of the following "cases," sketch a typical phase portrait, and write a short analysis of the evolution of the resulting relationship.

   a.   If each lover is disconcerted by the other's emotions ($b < 0$, $c < 0$) but is excited by his or her own feelings ($a > 0$, $d > 0$), what would be the course of their love affair? Pick some values for the parameters and find out.

   $$\dot{x} =$$
   $$\dot{y} =$$

   b.   Suppose Juliet's love is constant ($\dot{y} = 0$) no matter what Romeo feels. What would the system of equations look like? Examine some cases.

   $$\dot{x} =$$
   $$\dot{y} =$$

   c.   Do likes attract? Suppose the two lovers had exactly the same emotional profile in terms of their response to each other and their response to their own feelings. Investigate some situations and write a short analysis.

   $$\dot{x} = ax + by$$
   $$\dot{y} = bx + ay$$

   d.   Suppose Romeo and Juliet are both enthusiastic ($a, b, c, d > 0$) or both cautious ($a, b, c, d < 0$). What is the course of their love affair? Suppose one is cautious and one is enthusiastic? It is interesting to compare using the same initial conditions.

   both enthusiastic:

   $$\dot{x} =$$
   $$\dot{y} =$$
   both cautious:

   $$\dot{x} =$$
   $$\dot{y} =$$
   one cautious, one enthusiastic:

   $$\dot{x} =$$
   $$\dot{y} =$$

References

[MF] McDill, J.M. and Bjørn Felsager. "The Lighter Side of Differential Equations," *College Mathematics Journal* 25 (November 1994): 448–452.

[SS1] Strogatz, Steven. "Love Affairs and Differential Equations" *Mathematics Magazine* 61 (February 1988): 35.

[SS2] Strogatz, Steven. *Nonlinear Dynamics and Chaos, with Applications to Physics, Biology, Chemistry and Engineering.* Reading: Addison-Wesley, 1994.

Lab 18: Tool Instructions

Romeo and Juliet Tool

Setting Initial Conditions

Click the mouse on the xy graphing plane to set the initial conditions for a trajectory.

Clicking while a trajectory is being drawn will start a new trajectory.

Matrix Element Values

Click the arrow buttons to the left and right of the matrix elements to increase and decrease their values respectively.

Buttons

Click the mouse on the [**Draw Field**] button to draw a grid of vectors over the xy-plane.

Click the [**Pause**] button to stop a trajectory without canceling it.

Click the [**Continue**] button to resume the motion of a paused trajectory.

Click the mouse on the [**Clear**] button to remove all trajectories and vectors from the graphs.

19

The Glider

Tools Used in Lab 19

The Glider

If you've ever played with a balsa-wood glider, you know that it flies in a wavy path if you throw it gently, and it does loop-the-loops if you throw it hard. How is this all explained by phase plane analysis?

1. Basic Glider Flight

The motion of a glider is approximately governed by the dimensionless equations

$$\frac{dv}{dt} = -\sin\theta - Dv^2$$

$$\frac{d\theta}{dt} = \frac{v^2 - \cos\theta}{v}$$

where $v > 0$ is the speed of the glider, and θ is the angle that its nose makes with the horizontal. The angle θ is zero when the plane is flying level with the ground.

These equations capture the important forces: $\sin\theta$ is the component of the gravitational force that tends to change the speed of the plane, and Dv^2 is the drag force, due to air resistance, that tends to slow the plane down; $\cos\theta$ is the orthogonal component of the gravitational force, tending to rotate the plane, and v^2 is the lift, tending to increase the angle, that allows the plane to fly. For further discussion of the physics behind this model, see Section 3.

Since these equations are nonlinear, there is no hope of solving them analytically, but we can *see* what will happen by plotting the path of the glider.

If you move the cursor onto the phase plane, the screen also shows the initial value of a quantity called E, defined as $E = v^3 - 3v\cos\theta$; you'll see why E is significant when you play with **The Glider** tool.

1.1   Find a spot in the θv plane with a positive value of E and click to start a trajectory. Sketch the resulting glider path.

        Repeat for a negative value of E:

In Exercises **1.2**–**1.5**, you should assume that $D = 0$. (This is the simplest case.)

1.2   When the glider is in steady, level flight, what are the values of θ and v?

1.3   What is the most negative value of E that the glider can have?

1.4   Show that as the glider flies, the value of E stays constant, even though θ and v keep changing. (*Hint:* Calculate dE/dt and remember to use the chain rule correctly.)

1.5   For some initial conditions, the glider does loop-the-loops, while for others, it flies along a wavy path that is almost level. How can you tell ahead of time which type of behavior will occur?

2. Adding Drag

The parameter D measures the strength of the drag on the plane due to air resistance. It is initially set to 0, but you can change it with a slider. For the following exercises use a drag coefficient $D > 0$.

2.1   What is different about the flight path when $D > 0$, as compared to $D = 0$?

2.2   If you have already learned about classification of fixed points, find and classify all the fixed points of the system.

2.3   What is special about the value $D_c = 2\sqrt{2} \approx 2.82$ marked on the slider?

3. A Deeper Look at the Physics

For mathematical convenience, and to reduce the number of parameters, the equations discussed in Sections 1 and 2 were a dimensionless version of the true equations of motion. Now we consider the full equations, with all the units included, so that we can understand the physics in more detail. If we write Newton's law $F = ma$ in a moving coordinate system, with one axis tangent to the flight path, and another axis normal to it, the governing equations become

$$m\frac{dV}{dT} = -\text{drag} - mg\sin\theta$$

$$mV\frac{d\theta}{dT} = \text{lift} - mg\cos\theta$$

where m is the mass of the glider, g is the acceleration due to gravity, $V > 0$ is the speed of the glider, and θ is the angle that its nose makes with the horizontal.

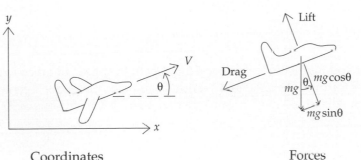

Coordinates   Forces

Let's try to understand the various forces. The terms $mg\sin\theta$ and $mg\cos\theta$ are the tangential and normal components of the gravitational force. The term $mV\frac{d\theta}{dT}$ in the normal-force equation looks strange at first, but it is simply the centripetal force. This is clear if we realize that $\frac{d\theta}{dT} = \frac{V}{R}$, where R is the instantaneous radius of curvature; then $mV\frac{d\theta}{dT} = mV^2/R$, a familiar expression for the centripetal force. The drag is a tangential force due to air resistance that tends to slow the plane down, whereas the lift is the normal-force that allows the plane to fly. Experiments indicate that both the drag and lift are proportional to V^2, to a good approximation, so the equations become

$$m\frac{dV}{dT} = -bV^2 - mg\sin\theta$$

$$mV\frac{d\theta}{dT} = cV^2 - mg\cos\theta$$

where b and c are constants.

3.1   Show that after appropriate rescaling of time and velocity, these equations can be reduced to the equations considered earlier.

3.2   Give a physical interpretation of the characteristic velocity V_c.

3.3   Given $V(T)$ and $\theta(T)$, how could you figure out the glider's flight path in space? In other words, if x and y are Cartesian coordinates for the horizontal and vertical location of the glider's center of mass, what are the equations that you would need to solve to find x and y as functions of T?

3.4   For the case of no drag ($D = 0$), show that the equations for the system imply that the total energy of the glider is conserved. (Please derive this directly from the equations of motion, not just from some physical reasoning.)

3.5   We found earlier that in the absence of drag, the quantity $E = v^3 - 3v\cos\theta$ remains constant on any given flight path. Is this equivalent to saying that energy is conserved when there is no drag?

4. For Further Exploration

The model discussed here is known as Lanchester's "phugoid theory," proposed in 1908—it is one of the the earliest theories for the flight path of an aircraft. For more about this model, as well as other aspects of flight dynamics, see von Mises (1959), pp. 539–545 or Miele (1962), pp. 271–273.

References

Lanchester, F. W.. *Aerodonetics*. London, 1908.

Miele, Angelo. *Flight Mechanics*. Vol. I, *Theory of Flight Paths*. Reading, MA: Addison-Wesley, 1962.

von Mises, Richard. *Theory of Flight*. New York: Dover, 1959.

Lab 19: Tool Instructions

The Glider Tool

Setting Initial Conditions

Click the mouse on the θV plane to set the initial conditions for a trajectory.
Clicking while a trajectory is being drawn will start a new trajectory.

Parameter Sliders

Use the slider to set the drag constant D.
Press the mouse down on the slider knob and drag the mouse back and forth, or click the mouse in the slider channel at the desired value for the parameter.

Buttons

Click the [**Pause**] button to stop a trajectory without canceling it.
Click the [**Continue**] button to resume the motion of a paused trajectory.
Click the mouse on the [**Clear**] button to remove all trajectories and vectors from the graph.

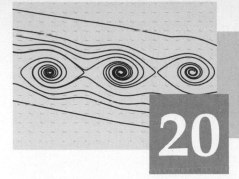

Nonlinear Oscillators: Free Response

20

Tools Used in Lab 20

Pendulums

The child on the backyard swing wants you to start her as high as you can manage with as much initial angular velocity as you can muster. Assumptions about small angular displacements no longer hold. Is there a loop-the-loop in her future?

1. The Undamped Pendulum

The equation for the free (unforced) pendulum without damping is

$$\frac{d^2\theta}{dt} + \frac{g}{L}\sin\theta = 0,\qquad(1)$$

where g is the acceleration of gravity, θ, a function of time t, is the angular displacement from the downward vertical, and L is the length of the pendulum. We define $\omega \equiv \sqrt{g/L}$ to simplify Equation (1), which becomes

$$\frac{d^2\theta}{dt} + \omega^2\sin\theta = 0.\qquad(2)$$

1.1 The Linear Simplification

For small values of θ, the values for θ and $\sin\theta$ radians are nearly equal. Use this fact to simplify Equation (1). Then solve the new linear equation to obtain the formula for the resulting simple harmonic motion. Show that the frequency in radians per second is the frequency ω defined above.

This motion is illustrated in the **Pendulums** tool in the **Linear** option. Now that you have reviewed the linear model, move on to the real thing, the nonlinear pendulum. Select the undamped, unforced **Simple Nonlinear** option. Select some nontrivial initial conditions and observe the resulting phase plane trajectories and time series plots for the angle, θ, and the angular velocity, $\dot{\theta}$.

1.2   The Phase Plane Portrait for the Free Undamped Pendulum

In your responses to the following questions on the pendulum motions that produced the phase plane below, include observations about closed orbits, simple harmonic motion, unstable and/or stable equilibria, and, of course, loop-the-loops or rotations.

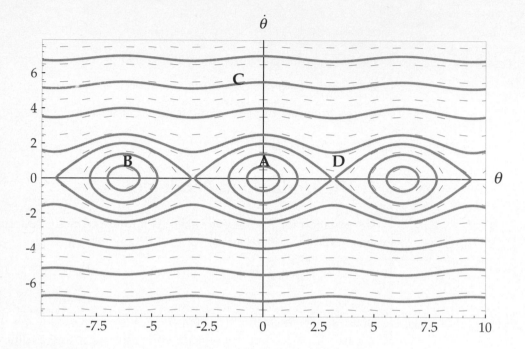

a.   Discuss the motion of the pendulum that results in trajectory A.

b.   Orbits A and B have the same size and shape; both orbits are the smallest shown and are horizontal translates of each other. Discuss the motion of the pendulum that results in trajectory B. How does it differ from trajectory A, if at all?

c.   Discuss the motion of the pendulum that results in trajectory C.

d.   Describe what happens to the pendulum at the saddle near D.

2. The Damped Unforced Pendulum

If we include viscous damping with damping constant b, the equation becomes

$$\ddot{\theta} + b\dot{\theta} + \omega^2 \sin\theta = 0 \qquad\qquad (3)$$

2.1   Rewrite Equation (3) as a linear system of first-order differential equations for angular velocity, $\dot{\theta} = \mu$ and angular acceleration, $\dot{\mu}$.

$$\dot{\theta} = \mu$$

$$\dot{\mu} =$$

2.2   A phase plane diagram for the damped pendulum is shown for $b = 0.2$ and $\omega = 1$.

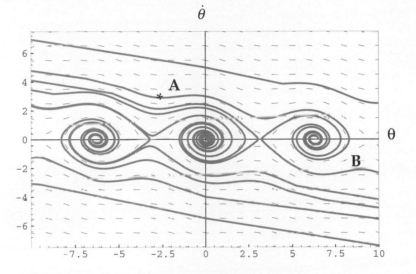

a.   Carefully describe the motion of the pendulum that corresponds to trajectory A in the preceding graph. Assume that A signifies the beginning of the trajectory at time $t = 0$. Be sure to start with your estimation of the signs of initial conditions $\theta(0)$ and $\dot{\theta}(0)$.

b.   Make the same sort of analysis of trajectory B. What is happening to the pendulum?

3. Further Exploration

3.1   Take a look at the **Damped Nonlinear** and **Forced Damped Pendulum** options, and describe what is different about the behaviors in each case. This is studied further in Lab 26, Chaos in Forced Nonlinear Oscillators.

Lab 20: Tool Instructions

Pendulums Tool

Setting Initial Conditions

Click the mouse on the $\theta \dot\theta$ plane to set the initial conditions for a trajectory.
Clicking while a trajectory is being drawn will start a new trajectory.

Time Series Buttons

The buttons labeled
 [ ] **position**
 [ ] **velocity**
 [ ] **acceleration**
toggle the time series graphs on and off.

Other Buttons

Click the mouse on [**Linear (SHO)**], [**Simple Nonlinear**], [**Damped Nonlinear**], or [**Forced Damped**] buttons to select a pendulum model.
Click the [**Pause**] button to stop a trajectory without canceling it.
Click the [**Continue**] button to resume the motion of a paused trajectory.
Click the mouse on the [**Clear**] button to remove all trajectories and vectors from the graph.

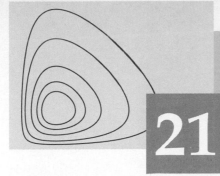

Predator-Prey Population Models

21

Tools Used in Lab 21

Hudson Bay Data (Hare-Lynx)
Lotka-Volterra
Lotka-Volterra with Harvest

How can we model the interaction between a species of predators and their prey, a species of herbivores? Using the well-known Lotka-Volterra Predator-Prey system, we explore the effects of the various constants and the effect of harvesting on the numbers of each species.

1. Basic Model

The **Lotka-Volterra** tool gives the predator-prey equations in the form

$$\frac{dH}{dt} = aH - bHP$$
$$\frac{dP}{dt} = -cP + d(bHP) \tag{1}$$

where H and P are, respectively, the prey (herbivore) and predator populations.

During World War I, the biologist D'Ancona wondered why the decrease in fishing in the Mediterranean Sea brought on by the war caused an increase in the percentage of the catch that was shark. The mathematician Vito Volterra used the preceding model to provide the answer.

The Lotka-Volterra equations have since been applied to many similar problems. The **Hudson Bay Data (Hare-Lynx)** tool displays data on kills of lynx and hares in the Arctic over a number of years. Compare the graphs given in the **Hudson Bay Data (Hare-Lynx)** tool to those predicted by the model in the **Lotka-Volterra** tool.

Use the **Lotka-Volterra** tool to answer the following questions.

1.1 Explain the biological meaning of each parameter.

  a:

  b:

c:

d:

1.2   In terms of the constants, what is the value of the equilibrium (H_E, P_E) for the system?

1.3   In the absence of predators, what happens to the population of herbivores? Why? Is this reasonable?

1.4   In the absence of herbivores, what happens to the population of predators? Why? Is this reasonable?

1.5   Experiment with the **Lotka-Volterra** tool to find the effect of changing each parameter and describe the effect in words.

2. Additional Exercise

In the Lotka-Volterra equation, take $a = 1.10$, $b = 0.90$, $c = 1.00$, $d = 1.00$. Set

$$\frac{dP}{dH} = \frac{dP/dt}{dH/dt} = \frac{H(1.10 - 0.90P)}{P(0.90H - 1.00)}$$

Use separation of variables followed by an appropriate substitution to find a function $f(H,P)$ so that the solution curves of the system lie on the level curves of the function f and we can conclude that the solution curves are indeed closed curves. Finding an argument for the conclusion is challenging!

3. Adding Harvesting

What happens if you harvest one or both species?

With harvesting, the Lotka-Volterra equations become

$$\frac{dH}{dt} = aH - bHP - h_h H$$
$$\frac{dP}{dt} = -cP + d(bHP) - h_p P \tag{2}$$

Use the **Lotka-Volterra with Harvest** tool to answer the following questions.

3.1   Explain the meaning of each parameter.

h_h:

h_p:

3.2   In terms of the constants, what is the value of the equilibrium (H_E, P_L) when there is harvesting?

3.3   What is the effect of harvesting herbivores? Does the model allow the possibility of extinction of the herbivores with harvesting?

3.4   What is the effect of harvesting predators? Does the model allow for the extinction of the predators with harvesting?

3.5   What is the effect of harvesting both species equally? Do both species become extinct? If not, who wins? When is pest control a bad idea?

Lab 21: Tool Instructions

Hudson Bay Data (Hare-Lynx) Tool

This tool displays data on hare and lynx populations from the Hudson-Bay Trading Company. It is not meant to be interactive.

Lotka-Volterra Tool

Setting Initial Conditions

The initial conditions for the active trajectory are displayed to the right of the graphing plane. They are set using either the mouse or the keyboard.

1. Click the mouse on the HP graph to set $H(0)$ and $P(0)$.
2. Click the mouse on the value for one of the initial conditions displayed beside the HP graph to activate a keyboard editor. Set a new number using the number keys, the right and left arrow keys, and the Delete key. Press the **[Return]** key or click the mouse away from the number to leave the editor, set the initial value, and start the trajectory.
3. Clicking while a trajectory is being drawn will stop the trajectory.

Parameter Sliders

Use the sliders to set the growth rate a, the predation rate b, the predator mortality rate c, and the food conversion rate d.

Press the mouse down on the slider knob for the parameter you want to change and drag the mouse back and forth, or click the mouse in the slider channel at the desired value for the parameter.

Buttons

Click the mouse on the **[Clear]** button to remove all output from the graphing plane.
Click the mouse on the **[Pause]** button to stop a trajectory without canceling it.
Click the mouse on the **[Continue]** button to resume the motion of the paused trajectory.

Lotka-Volterra with Harvest Tool

Setting Initial Conditions

The initial conditions for the active trajectory are displayed to the right of the graphing plane. They are set using either the mouse or the keyboard.

1. Click the mouse on the HP graphing plane to set $H(0)$ and $P(0)$.
2. Click the mouse on the value for one of the initial conditions displayed beside the HP graph to activate a keyboard editor. Set a new number using the number keys, the right and left arrow keys, and the **[Delete]** key. Press the **[Return]** key or click the mouse away from the number to leave the editor, set the initial value, and start the trajectory.
3. Clicking while a trajectory is being drawn will stop the trajectory.

Parameter Sliders

Use the sliders to set the growth rate a, the predation rate b, the predator mortality rate c, and the food conversion rate d, the harvest rate for herbivores h_H, and the harvest rate for predators h_P.

Press the mouse down on the slider knob for the parameter you want to change and drag the mouse back and forth, or click the mouse in the slider channel at the desired value for the parameter.

Buttons

Click the mouse on the **[Clear]** button to remove all output from the graphs.
Click the mouse on the **[Pause]** button to stop a trajectory without canceling it.
Click the mouse on the **[Continue]** button to resume the motion of the paused trajectory.

Competing Species Population Models

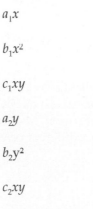

Tools Used in Lab 22

Competitive Exclusion

Two species, x and y, are competing for the same resources. Given initial populations, the carrying capacities of the habitat, the intrinsic growth rates, and the competitiveness of each species, what are the possible outcomes? Which parameters are most important in determining an outcome? Can an outcome be predicted?

1. The Competition Model

Competition between two species with populations x and y is typically described algebraically by equations of the form

$$\frac{dx}{dt} = x(a_1 - b_1 x - c_1 y)$$

$$\frac{dy}{dt} = y(a_2 - b_2 y - c_2 x). \tag{1}$$

1.1 Explain the relationship of each of the six terms to the populations and their interaction.

$a_1 x$

$b_1 x^2$

$c_1 xy$

$a_2 y$

$b_2 y^2$

$c_2 xy$

The **Competitive Exclusion** tool gives the equations in a form that provides a clearer geometric indication of the meanings of the constants. This is *not* an obvious fact, but it will become clear in Exercise **1.3**.

$$\frac{dN_1}{dt} = r_1 N_1 \frac{K_1 - N_1 - B_1 N_2}{K_1}$$

$$\frac{dN_2}{dt} = r_2 N_2 \frac{K_2 - N_2 - B_2 N_1}{K_2} \tag{2}$$

1.2   Show that the equations in Equation (2) are equivalent to the equations in Equation (1).

1.3   Experiment with the constants in Equations (2) and describe the effect of changing each.

K_1

K_2

B_1

B_2

r_1

r_2

2. Possible Behaviors of Solutions

2.1   As shown below, there are four possible nullcline configurations in the first quadrant (h for horizontal and v for vertical). In each region carved out by the nullclines, draw arrows to indicate the general direction of the solutions in that region. Mark on each sketch all stable equilibria as • and all unstable equilibria as ∘ . Finally, in a different color, sketch some typical phase plane trajectories for each of these cases.

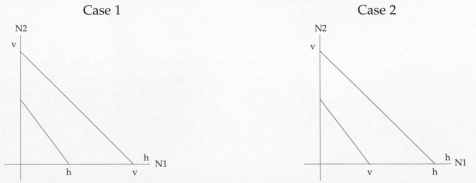

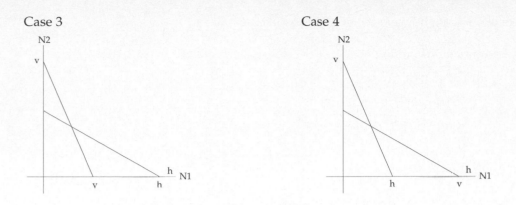

Case 3

Case 4

2.2   For each case, sketch a typical time series in two colors, one color for N_1, another for N_2.

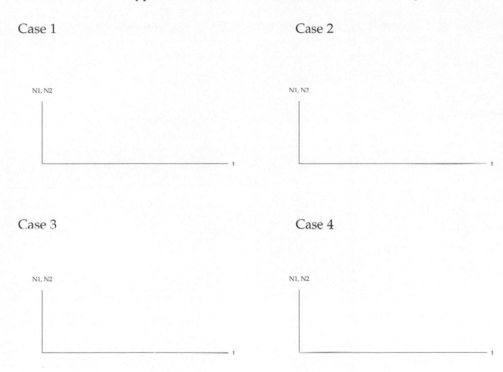

Case 1

Case 2

Case 3

Case 4

3. Connecting with the Biology

3.1   Select the best case for each of the situations described below and justify your choice.

a.   You are working with a regional wetlands ecological commission to save frogs, of which N_1 and N_2 are populations of two species.

b.   You are working for a pest control company to eradicate nasty species N_1 (you name it).

3.2   If you are in a situation where the graph looks like Case 3:

   a.   What parameters would you change to achieve the goal in Exercise **3.1b**?

   b.   Give a possible "real life" scenario that describes how you might change the environment to accomplish this goal.

Lab 22: Tool Instructions

Competitive Exclusion Tool

Setting Initial Conditions

The initial conditions for the active trajectory are displayed to the right of the graphing plane. They are set using either the mouse or the keyboard.

1. Click the mouse on the $N_1 N_2$ graphing plane to set $N1(0)$ and $N2(0)$.
2. Click the mouse on the value for one of the initial conditions displayed beside the $N_1 N_2$ graph to activate a keyboard editor. Set a new number using the number keys, the right and left arrow keys, and the **[Delete]** key. Press the **[Return]** key or click the mouse away from the number to leave the editor, set the initial value, and start a trajectory.
3. Clicking while a trajectory is being drawn will cancel the trajectory.

Parameter Sliders

The constants for carrying capacity K, competitiveness B, and growth rate r, for each species are set using the horizontal sliders on the right side of the screen.

Press the mouse down on the slider knob for the parameter you want to change and drag the mouse back and forth, or click the mouse in the slider channel at the desired value for the parameter.

Buttons

Click the mouse on the **[Clear]** button to remove all output from the graphs.

Click the mouse on the **[Pause]** button to stop a trajectory without canceling it.

Click the mouse on the **[Continue]** button to resume the motion of the paused trajectory.

Part

V Chaos and Bifurcation

23 Bifurcations in 1-D

Water suddenly freezes into ice as the temperature is lowered below the freezing point. This is an example of a bifurcation: a qualitative change in the behavior of a system as a parameter is varied.

1. Bifurcation

For an autonomous differential equation $x' = f(x,r)$ with one parameter r, changes in the behavior of the solutions due to a small change in r are usually gradual and continuous. At certain values of r, however, the tx pictures may exhibit a sudden *qualitative* change, which is called a **bifurcation.** This lab allows you to explore just how the gradual changes and drastic changes can occur simultaneously.

There are four common ways in which bifurcation occurs in a family of one-dimensional equations $x' = f(x,r)$, where r is a parameter. The four tools for this lab give one example for each case: **Saddle-Node Bifurcation, Transcritical Bifurcation, Pitchfork Bifurcation: Supercritical, Pitchfork Bifurcation: Subcritical.**

We will examine each of these in turn later in the lab, but for now we just want to show what bifurcation means in the tx-plane. The four tools all work the same way.

1.1   To see an example of bifurcation, open any one of the tools for this lab. By clicking with the mouse, draw some representative solutions in the tx-plane and sketch the results on the appropriate graph. Then repeat, using both a positive value for r and a negative value.

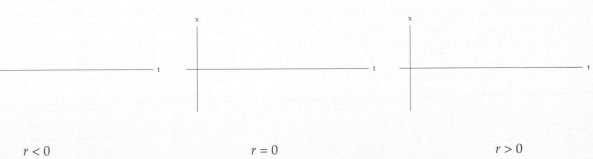

$r < 0$ $r = 0$ $r > 0$

Name of tool used: Equation:

1.2   Describe in words how the behavior of the solutions $x(t)$ has changed as r moves from left to right in Exercise **1.1**:

1.3   Repeat the experiment of Exercise **1.1** for a second tool:

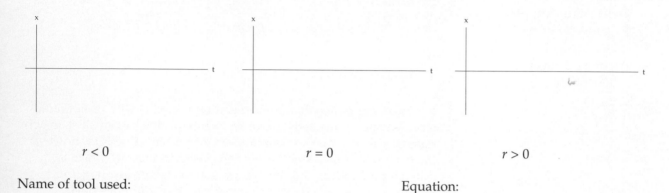

            $r < 0$ $r = 0$ $r > 0$

Name of tool used: Equation:

1.4   Describe in words how the behavior of the solutions $x(t)$ has changed as r moves from left to right in Exercise **1.3**:

1.5   Describe the differences in behavior between your two examples from Exercises **1.1** and **1.3**:

These first exercises should have alerted you to the phenomena we wish to study. Just what is it that changing r does? Before we move on to the details in each of the four types of bifurcation, we need to look at the three graphs in each tool and see how they relate to each other. The **phase line** is the key that ties them all together, and is the focus of the answer to the question we just asked. Although the phase line was discussed in Lab 3, Sections 3 and 4, most textbooks have not given it or the graphs in the xx'- and rx- planes the necessary prominence to serve as a foundation for analyzing bifurcations.

2. The Phase Line and its Role in *xx′* and *rx* Graphs

Once a value of r has been chosen, the xx' graph on the upper left can be drawn from the differential equation; the phase line forms its horizontal axis. The points at which the graph crosses the x-axis are equilibrium points, where $x' = 0$. The points where x' is positive are where solutions are moving in the direction of increasing x; where x' is negative, solutions are moving in the direction of decreasing x—this information determines the directions of the arrowheads on the phase line.

The same phase line is rotated to a vertical position in the tx graph in the lower left of the screen.

2.1   Using any of the four tools, sketch an xx' graph and the corresponding tx graph, with some solutions.

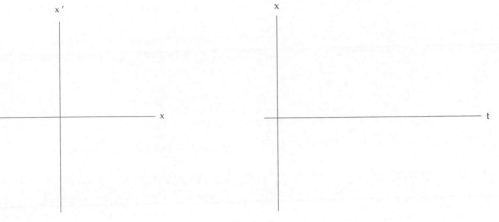

2.2   Mark the phase line in the graphs for Exercise **2.1**. It is the horizontal axis on the xx' graph and the vertical axis on the tx graph. These should be marked with exactly the same configuration in the x direction in each case, with marked equilibrium points and appropriate arrows between. The convention is to color an equilibrium point solid if it is *stable* (that is, if arrowheads on either side are coming toward it), to leave it open if the equilibrium is *unstable* (both arrowheads point away).

2.3   How might an equilibrium be neither stable nor unstable? Support your answer with a picture of a possible case.

2.4.   What behaviors on the tx graph correspond to the points of equilibrium on the phase line? to the arrows pointing up on the phase line? to the arrows pointing down?

2.5   To emphasize that the qualitative features of the *tx* graph can be inferred from the *xx'* graph, draw a *tx* graph for each of the following *xx'* graphs. Start by drawing the horizontal phase line on the *xx'* graph, then the same thing vertically on the *tx* graph; then sketch some sample solutions from the information contained in the phase line. Pay attention to whether a given solution *x(t)* is increasing, decreasing, or constant. Try also to get the proper concavity: if a solution is rising, is it getting steeper as it rises (accelerating) or is it getting shallower (decelerating)?

a.

b.

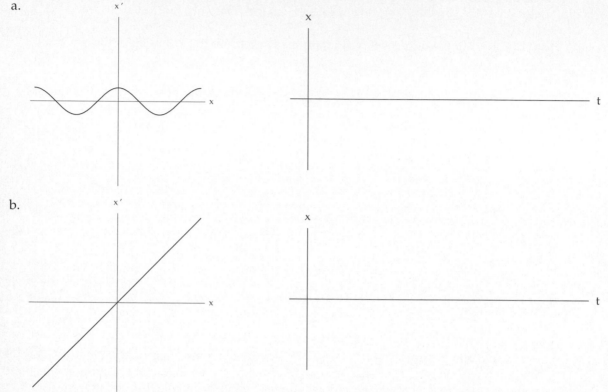

The third graph in each tool is composed of adjacent vertical phase lines. That is, on the *rx* graph, for each value of *r* you could plot a phase line vertically. The locus followed by a stable equilibrium as it moves when *r* changes is drawn as a solid curve; the locus followed by an unstable equilibrium as it moves when *r* changes is drawn as a dotted curve. The result is called a **bifurcation diagram** because it summarizes for any equation $x' = f(x,r)$ what you can expect for a phase line (and hence for the behavior of solutions in the *tx* plane) for any *r*.

2.6   For any of the tools, sketch here the *rx* bifurcation diagram.

Now you will demonstrate how to obtain key aspects of the other graphs from this one, without the computer! Without using the tool further, choose a value of *r* other than that where your example is set, for instance *r* = −1, and sketch by hand the critical features of the *xx'* graph—not from the equation, but just from the phase line information you can read from the *rx* graph at that *r* value. That is, lay along the horizontal axis of the *xx'* graph the phase line information with appropriate arrows, since you know which equilibrium is stable and which is unstable by whether it lies on a solid or dotted bifurcation curve. Where the arrows point to the right, sketch *x'* above the horizontal axis, and where they point to the left, sketch *x'* below the axis.

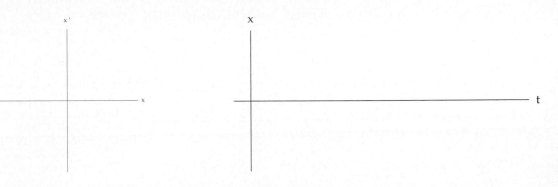

This should give a very good qualitative sketch, and from that you can sketch a *tx* graph.

2.7   Return to the computer tool, input your new *r* value, and sketch below the *xx'* and *tx* graphs that result. You can expect that the details of how high the computer *xx'* graph rises above or below the horizontal axis will not usually match your hand sketch of the same graph, but the qualitative features of where the graph crosses the axis, and how, should be exactly the same.

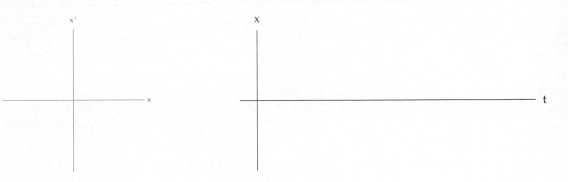

Explain the differences that may occur between your computer and hand sketches of the *tx* graph:

At this point you should have a pretty good understanding of all the graphs in each tool and how they relate to one another. It is time to look individually at each of the four kinds of bifurcation for a one-dimensional differential equation, keeping in mind that the overriding question is "What does changing *r* do to the fixed points and the phase line?"

3. Saddle-Node Bifurcation

The special feature of saddle-node bifurcation is that as r changes in one direction or the other, the number of equilibria changes—you should note the number and the type in each example.

The name "saddle-node" comes from the two-dimensional analog, but you can think of "saddle" as an unstable fixed point, and "node" as a stable one.

3.1   A nice standard example of a saddle-node bifurcation is given by

$$\dot{x} = r + x^2. \qquad\qquad (1)$$

Open the **Saddle-Node Bifurcation** tool and experiment with different values of r, if you haven't already, to confirm this definition. Sketch the rx bifurcation diagram, and sketch a typical set of behaviors for a negative, zero, and positive value of r. Explain what happens to the equilibria on the phase line as r moves from left to right.

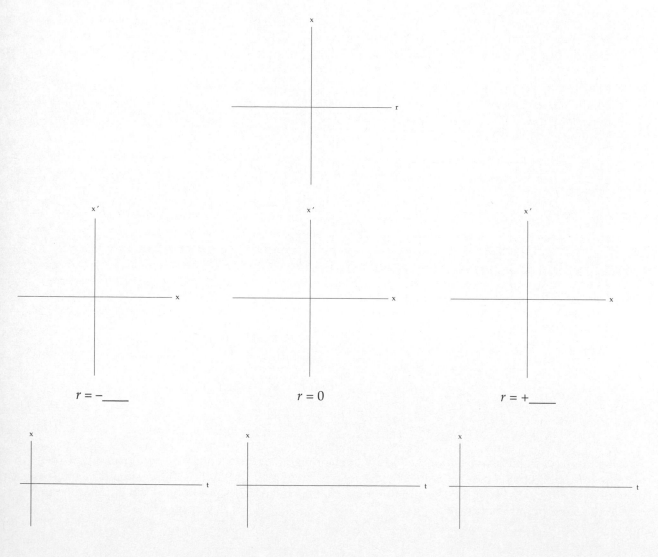

Explanation:

3.2   A nice example of a saddle-node bifurcation comes from tipping a wire that forms a well. Imagine a bead on this wire (and to be physically correct, you should imagine the whole apparatus in a vat of viscous goo—this provides viscous damping, so that whenever possible, the bead would tend to settle in the bottom of the well).

In the topmost curve, the black circle represents a stable equilibrium for the bead; whenever it is perturbed a bit, it will settle back down to the bottom of the well. The open circle represents another equilibrium; if the bead were perfectly balanced on this peak, it would stay there, but if it's even slightly perturbed, it will roll downhill away from the peak—hence this equilibrium is unstable. The arrows show the directions a bead would slide on each segment of the wire.

For each of the subsequent positions of the wire, draw the location of the stable and unstable equilibria—the bottom of the well and the top of the peak from the point of view of gravity directed straight down the page. Color the stable equilibrium black and leave the unstable one open. Draw the arrows as well. What happens?

↓

force of
gravity

Vat of viscous goo,
for tipping wire well

3.3   For $x' = r - x - e^{-x}$, find where the bifurcation is located, and sketch representative xx', rx, and tx graphs. This should be done on a separate sheet of paper. *Hint*: Since the xx' graph is awkward to draw by hand, an easy trick is to graph separately $r - x$ and e^{-x} on the same axes; the intersections show the points (r,x) where the equilibria are located. You can plot this for several different r values, and then make rx and tx graphs from the information you can glean. This exercise can be done qualitatively by hand, or more precisely using open-ended graphing tools.

4. Transcritical Bifurcation

In a transcritical bifurcation, the key is that two equilibria change stability. They don't disappear.

4.1   The standard transcritical bifurcation generally encountered is provided by

$$\dot{x} = rx - x^2. \tag{2}$$

Open the **Transcritical Bifurcation** tool, sketch the rx bifurcation diagram, and experiment with different values of r. Explain, with sketches of the xx' plane and the corresponding tx sketches, how it is that the two equilibria exchanged stabilities.

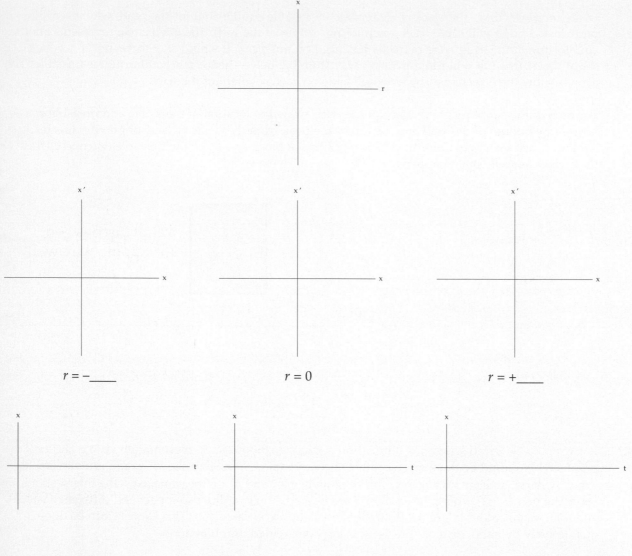

$r = -\underline{\qquad}$ $r = 0$ $r = +\underline{\qquad}$

Explanation:

4.2   Locate the bifurcation value for r in the differential equation $x' = rx - x(1-x)$ on a separate sheet of paper. By hand or with open-ended computer tools, sketch the xx', rx, and typical tx graphs.

5. Pitchfork Bifurcation

A pitchfork bifurcation arises when certain symmetries exist; its bifurcation diagram in the rx plane looks like a pitchfork, with three tines and a single handle. Although a convention has arisen to distinguish two subtypes of pitchfork bifurcation, supercritical and subcritical, some general observations can be made about the pitchforks in both cases.

5.1   From the tools **Pitchfork Bifurcations: Supercritical** and **Pitchfork Bifurcations: Subcritical**, sketch the rx bifurcation diagrams.

Supercritical Pitchfork
$$x' = rx - x^3$$

Subcritical Pitchfork
$$x' = rx + x^3$$

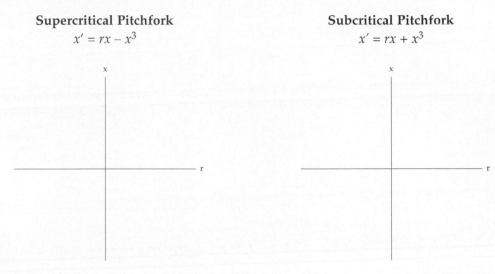

5.2   What happens to the stability of the handle when it becomes the middle tine of the pitchfork?

5.3   The outer two tines of the pitchfork represent a symmetric pair of equilibria on one side or the other of the bifurcation value. What kinds of stability do they represent?

5.4   Describe the differences between supercritical and subcritical pitchfork bifurcations.

5.5   For each of the two types of pitchfork bifurcations, from the *rx* diagrams in **Exercise 5.1**, predict, by hand, the *tx* behaviors for the various cases.

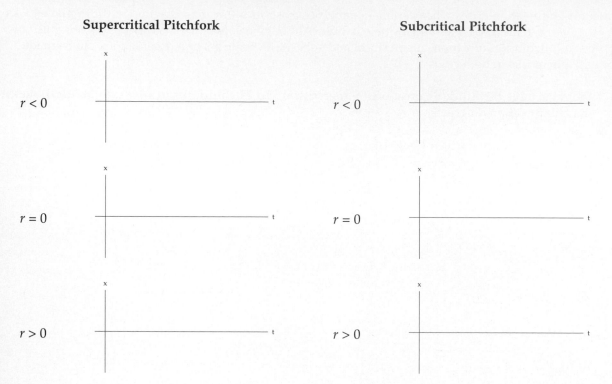

Supercritical Pitchfork   **Subcritical Pitchfork**

$r < 0$ $r < 0$

$r = 0$ $r = 0$

$r > 0$ $r > 0$

5.6   With the **Pitchfork Bifurcation: Supercritical** and **Pitchfork Bifurcation: Subcritical** tools, confirm your sketches in Exercise 5.5. Explain any differences, and distinguish between the two subtypes.

5.7   On a separate sheet of paper, locate the bifurcation value(s) for *r* in the equation $x' = rx + x/(1+x^2)$. Sketch xx', rx, and typical tx graphs.

6. Simple Physical Examples

One very simple example of bifurcation is the shape of a thin plastic coffee stirrer being squeezed when held by its ends between thumb and forefinger. Eventually the stirrer will buckle, to one side or the other. In a symmetrical situation, buckling to either side is equally likely, so this is a nice example of a pitchfork bifurcation.

Squeeze!

6.1   Would this be a subcritical or supercritical pitchfork bifurcation? Why?

Population models are another source of bifurcation behavior. A good example is provided by the harvesting example of Lab 3 on single species population models.

6.2   Open the **Logistics with Harvest** tool of Lab 3, Section 4. Here $x' = x(1-x) - h$. At what value of h does bifurcation occur?

6.3   What is the nature of the bifurcation of Exercise **6.2** when the harvesting becomes too intense? Name it mathematically, but describe it biologically.

6.4   We have not readily thought up a simple example of a transcritical bifurcation. Can you? We would be happy to use the best submitted to us in our next edition, and give you credit.

7. Additional Exercises

You can now try to find and identify bifurcations from equations in general with a parameter.

7.1   Make a chart that compares and classifies the three basic types of bifurcation for one-dimensional equations $x' = f(x,r)$, based on your observations in Sections 3, 4, and 5. What properties are special to each type? How might you sum up the identifying criteria for each type?

	Saddle-Node	Transcritical	Pitchfork
How do xx' graphs change as r changes?			
How do tx graphs change as r changes?			
Draw a typical rx graph, summarizing all the phase lines.			
Identifying criteria:			

7.2   For the following equations, identify which kinds of bifurcations can occur, and where. ("None" is a possible correct answer.) You can do it by hand or by using a computer with open-ended graphing tools, and you can work from whatever information you find easiest to get in each case. Clearly describe (on a separate page) your steps and your reasoning. In some cases you may find more than one bifurcation, and they might be of different types; in some cases you may find additional phenomena to watch out for or comment upon.

a. $x' = r + \cos(x)$ c. $x' = rx \sin x$, for $r > 0$

b. $x' = r + x/2 - x/(1 + x)$ d. $x' = rx - \sin x$

Lab 23: Tool Instructions

Saddle-Node Bifurcation Tool

Setting Initial Conditions
Click the mouse on the $x\dot{x}$ graph on the top left or the tx graphing plane on the lower left to see graphical output.
Clicking while a trajectory is being drawn will start a new trajectory.

Parameter Sliders
Use the sliders to change the values for the parameter r.
Press the mouse down on the slider knob and drag the mouse back and forth, or click the mouse in the slider channel at the desired value for the parameter.

Buttons
Click the mouse on the **[Draw Field]** button to draw a slope field.
Click the mouse on the **[Clear]** button to remove all trajectories from the graph.

Transcritical Bifurcation Tool

Setting Initial Conditions
Click the mouse on the $x\dot{x}$ graph on the top left or the tx graph on the lower left to see graphical output.
Clicking while a trajectory is being drawn will start a new trajectory.

Parameter Sliders
Use the sliders to change the values for the parameter r.
Press the mouse down on the slider knob and drag the mouse back and forth, or click the mouse in the slider channel at the desired value for the parameter.

Buttons
Click the mouse on the **[Draw Field]** button to draw a slope field.
Click the mouse on the **[Clear]** button to remove all trajectories from the graph.

Pitchfork Bifurcation: Supercritical Tool

Setting Initial Conditions
Click the mouse on the $x\dot{x}$ graph on the top left or the tx graphing plane on the lower left to see graphical output.
Clicking while a trajectory is being drawn will start a new trajectory.

Parameter Sliders
Use the sliders to change the values for the parameter r.
Press the mouse down on the slider knob and drag the mouse back and forth, or click the mouse in the slider channel at the desired value for the parameter.

Buttons
Click the mouse on the **[Draw Field]** button to draw a slope field.
Click the mouse on the **[Clear]** button to remove all trajectories from the graphs.

Pitchfork Bifurcation: Subcritical Tool

Setting Initial Conditions

Click the mouse on the $x\dot{x}$ graph on the top left or the tx graphing plane on the lower left to see graphical output.

Clicking while a trajectory is being drawn will start a new trajectory.

Parameter Sliders

Use the sliders to change the values for the parameter r.

Press the mouse down on the slider knob and drag the mouse back and forth, or click the mouse in the slider channel at the desired value for the parameter.

Buttons

Click the mouse on the **[Draw Field]** button to draw a slope field.

Click the mouse on the **[Clear]** button to remove all trajectories from the graphs.

24 Spruce Budworm

The spruce budworm is a serious pest in eastern Canada, where it attacks the leaves of the balsam fir tree. When an outbreak occurs, the budworms can defoliate and kill most of the fir trees in the forest in about four years.

1. Introduction

The equation

$$\frac{dx}{dt} - rx\left(1 - \frac{x}{k}\right) - \frac{x^2}{1 + x^2} \qquad (1)$$

gives an idealized population model for the spuce budworm, first proposed and analyzed by Ludwig, Jones, and Holling (*J. Anim. Ecol.* 47 (1978), 315). The variables t and x should be thought of as time and the spruce budworm population, respectively. The first term on the right side models the growth of the spruce budworm population without predation as logistic. The constant parameters $k > 0$ and $r > 0$ should be thought of as the carrying capacity of the forest (maximum number of spruce budworms that can live in the forest) and the growth rate of the spruce budworm population, respectively, in the absence of predation. The second term models the effect of predation, chiefly by birds. For a more detailed discussion of the spruce budworm models, see Steven H. Strogatz, *Nonlinear Dynamics and Chaos*, Addison-Wesley, Reading, MA, 1994.

1.1   What does (i) $\frac{dx}{dt} > 0$, (ii) $\frac{dx}{dt} < 0$, (iii) $\frac{dx}{dt} = 0$ imply about the change of the spruce budworm population with time?

2. Time Series

Due to the continuity of the right-hand side of Equation (1), the sign of $\frac{dx}{dt}$ can change only when $\frac{dx}{dt} = 0$. This happens when $x = 0$, or when

$$r\left(1 - \frac{x}{k}\right) = \frac{x}{1 + x^2} \qquad (2)$$

Note that the right-hand side of Equation (2) is independent of the parameters, and the left-hand side of

Equation (2) is linear in x. Now open the **Spruce Budworm: Time Series** tool. The two sides of Equation

(2), $y = r\left(1 - \frac{x}{k}\right)$ and $y = \frac{x}{1 + x^2}$, are plotted on the same graph. The parameters k and r can be varied

using the sliders.

2.1 How can you tell from the graphs of $y = r\left(1 - \frac{x}{k}\right)$ and $y = \frac{x}{1 + x^2}$ whether $\frac{dx}{dt}$ is positive, negative, or zero?

2.2 Find values of k and r for which the right side of Equation (1) has only one zero besides $x = 0$ and sketch the time series for your choice of parameters and representative choices of initial conditions.

2.3 For your choice of r and k in Exercise **2.2** and an initial value of spruce budworm population above the one positive equilibria of the differential equation in Exercise **2.2**, how does the spruce budworm population vary with time? Why?

2.4   Find values of k and r for which the right-side of Equation (1) has three zeros besides $x = 0$ and
sketch the time series for your choice of parameters and representative choices of initial conditions.

3. Another Viewpoint

Open the **Spruce Budworm: kr-Plane tool**. You choose values of the parameters k and r by moving the
cursor in the kr-plane. The graph of the right-hand side of from Equation (1) is displayed, as well as the
graphs of $y = r\left(1 - \dfrac{x}{k}\right)$ and $y = \dfrac{x}{1 + x^2}$,

3.1   What is the significance of the cusp in the kr-plane?

4. Bifurcation Diagram

The bifurcation diagram for Equation (1) is the set of points in krx-space for which $\dfrac{dx}{dt} = 0$. Since $x = 0$ is
an equilibrium, each point with $x = 0$ belongs to the bifurcation diagram. The rest of the bifurcation
diagram consists of those points satisfying Equation (2).

Open the **Spruce Budworm: rx-Plane** tool. Choose several values of k with the slider and observe the
graph of Equation (2) in the rx-plane. For k constant, the bifurcation diagram is this curve in the rx-plane.

4.1   Choose a value of k for which Equation (2) does not define x as a function of r (recall the vertical line test). Draw the bifurcation diagram for this fixed k (use a solid curve for those equilibria that are attractors and a dashed curve for those equilibria that are repellors). For a representative set of values of r, indicate the long-term behavior of the spruce budworm population x by drawing arrows on your bifurcation diagram (moving the cursor in the rx-plane chooses a value of r and displays the graph of $\dfrac{dx}{dt}$).

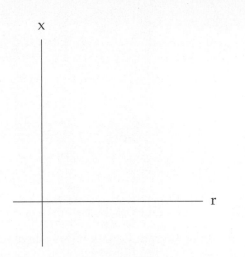

The graph of Equation (2) in krx-space can be generated from the graphs of Equation (2) in the rx-plane with k fixed by allowing k to vary. Open the **Spruce Budworm: Cusp** tool to view a graph of Equation (2). The interior of the cusp drawn in the kr-plane in the **Spruce Budworm: kr-Plane** tool, contains the points in the kr-plane covered by 3 points of the graph of Equation (2). That is, for each point (k, r) within the cusp there are 3 points on the graph of Equation (2) with those kr-coordinates.

5. Hysteresis

In modeling the spruce budworm population, we assume a fixed k and that the growth rate r of the spruce budworm population drifts slowly upward with time (because the forest cover increases) and then decreases as the spruce budworms kill off the trees. Open the **Spruce Budworm: Hysteresis** tool. Choose a value of the parameter k by clicking in the kr-plane. The software causes r to drift automatically. Observe how x varies with r (and thus with time) for various choices of k.

5.1   For the value of k you used in Exercise **4.1**, describe and explain this variation of spruce budworm population x with r and thus with time.

6. A Similar Equation

An equation that exhibits dynamics similar to the dynamics of the spruce budworm equation, but with simpler algebra is given by

$$\frac{dx}{dt} = a + rx - x^3 \qquad\qquad (3)$$

6.1 Fix $r = -1, 0, 1$ in Equation (3) and draw the bifurcation diagrams for the resulting three families of differential equations, each depending only on parameter a. *Hint:* Graph the function $a(x) = x^3 - rx$ and interchange the a-axis and the x-axis so that the x-axis is vertical and the a-axis is horizontal. Remember to dash portions of the curve corresponding to unstable equilibria.

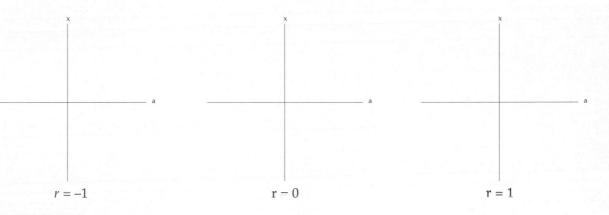

$r = -1$ $r = 0$ $r = 1$

You can use these bifurcation diagrams to visualize the entire bifurcation diagram of Equation (3), which consists of the graph of the surface $a + rx - x^3 = 0$ in rax-space.

6.2 If you project this surface onto the ra-plane, there is a cusp that consists of the points in the ra-plane that are images of exactly two points on the surface. Find the equation of this cusp and open the **Imperfect Bifurcation** tool to see this cusp.

6.3 Use the **Imperfect Bifurcation** tool to draw the bifurcation diagrams for the three families of differential equations, depending on the parameter r determined from Equation (3) by fixing $a = -1, 0, 1$. Note that when $a = -1$ or $a = 1$, the bifurcation diagram does not change very much with a small change in a but that when $a = 0$, the bifurcation diagram changes completely with any change in a, no matter how small.

Lab 24: Tool Instructions

Spruce Budworm: Time Series Tool

Setting Initial Conditions

Click the mouse on the graphing plane on the right (labeled tx) to see graphical output.
Clicking in the plane while a trajectory is being drawn will start a new trajectory.

Parameter Sliders

Use the sliders to change the values for the parameters r and k.
Press the mouse down on the slider knob for the parameter you want to change and drag the mouse up and down, or back and forth, or click the mouse in the slider channel at the desired value for the parameter.

Spruce Budworm: *kr*-Plane Tool

Setting Parameters

Move the mouse over the kr graph on the left to select values of r and k.

Spruce Budworm: *rx*-Plane Tool

Setting Initial Conditions

Move the mouse over the rx graph on the left to see the graphical output of dx/dt.

Parameter Sliders

Use the sliders to change the values for the parameter k.
Press the mouse down on the slider knob and drag the mouse back and forth, or click the mouse in the slider channel at the desired value for the parameter.

Spruce Budworm: Cusp Tool

This tool is demonstrative, not interactive.

Spruce Budworm: Hysteresis Tool

Setting Initial Conditions

Click the mouse on the kr graph on the left to select a value of k.

Imperfect Bifurcation Tool

Setting Initial Conditions

Click the mouse on the $x\dot{x}$ graph on the top left, the graphing plane of the lower left (labeled x), and the ra graphing plane on the top right to see the graphical output.

Clicking in the tx plane while the trajectory is being drawn will start a new trajectory.

Parameter Sliders

Use the sliders to change the values for the parameters a and r.

Press the mouse down on the slider knob for the parameter you want to change and drag the mouse up and down, or back and forth, or click the mouse in the slider channel at the desired value for the parameter.

Buttons

Click the mouse on the **[Draw Field]** button to draw a slope field.

Click the mouse on the **[Clear]** button to remove all trajectories from the graph.

25 Bifurcations in Planar Systems

Tools Used in Lab 25

Chemical Oscillator

2-D Saddle-Node
Bifurcation

Families of systems of first order autonomous equations, depending on a parameter, show surprising changes in behavior as the parameter is varied.

1. The Hopf Bifurcation

Lengyel, Rabai, and Epstein (*Journal of the American Chemical Society* (1990) 112, 9104) have posed and analyzed a particularly elegant model of an oscillating chemical reaction—the chlorine dioxide-iodine-malonic acid reaction—modeled by a system of two differential equations:

$$\frac{dx}{dt} = a - x - \frac{4xy}{1+x^2}$$
$$\frac{dy}{dt} = bx\left(1 - \frac{y}{1+x^2}\right)$$

(1)

The variables x and y should be thought of as modeling concentrations of I^- and ClO_2^-, while a and b should be thought of as chemical constants. Also see Steven H. Strogatz, *Nonlinear Dynamics and Chaos*. Addison-Wesley, Reading, MA: 1994.

As the concentrations of I^- and ClO_2^-, change, the color of the solution changes. Open the **Chemical Oscillator** tool for a simulation of the chemical reaction and for help answering the following questions. If you have a friend who is a chemist you might enjoy mixing up a batch and seeing whether the color changes indicated in the simulation are accurate.

For simplicity, we take $a = 10$ and let $b > 0$ be a parameter.

1.1   Find all equilibria of Equation (1).

1.2   Linearize Equation (1) at the equilibrium and find the eigenvalues.

1.3   Conclude that the equilibrium is a repellor if $b < b_c = 3.5$, and an attractor if $b > b_c$.

Let $A = (0,101)$, $B = (0,0)$, $C = (10,0)$, and $D = (10,101)$.

1.4   Show that the vector field points horizontally to the right along segment AB.

1.5   Show that the vector field points up along segment BC.

1.6   Show that the vector field points to the left along segment CD.

1.7   Show that the vector field points down along segment DA.

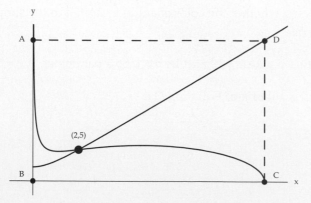

Let $\Re$ be the region bounded by the rectangle $ABCD$, with the point $(2,5)$ deleted. By **1.4**, **1.5**, **1.6**, and **1.7** no trajectory leaves the region bounded by the rectangle. Further, if $b < b_c$, the point $(2,5)$ is a repellor, so no trajectory can leave the region $\mathfrak{R}$. By the Poincaré-Bendixson theorem, the region $\mathfrak{R}$ contains a cycle. Use the **Chemical Oscillator** tool to convince yourself that this cycle is an attracting cycle that shrinks continuously to the equilibrium as b increases to b_c.

Equation (1) is said to have a **Hopf bifurcation** at b_c. That is, for $b > b_c$, (2,5) is an attracting equilibrium. At $b = b_c$ a bifurcation occurs. The equilibrium (2,5) becomes a repellor, and an attracting cycle is born that increases in size as b decreases from b_c.

2. Product Bifurcations

The Hopf bifurcation is intrinsically two dimensional. A lot of two-dimensional bifurcations are intrinsically one-dimensional. The simplest of these are the product bifurcations. Given a family of first-order equations, for example,

$$\dot{x} = r + x^2 , \qquad\qquad (2)$$

one can form the **product family** by adding the equation $\dot{y} = -y$:

$$\frac{dx}{dt} = r + x^2$$

$$\frac{dy}{dt} = -y \qquad\qquad (3)$$

The family Equation (2) has a 1-D saddle node bifurcation at $r = 0$ (See Lab 23). Knowing the behavior of the family Equation (2) makes it very easy to figure out the behavior of the product family Equation (3). Fix an interesting value of r. Take the x-axis as horizontal and draw the phase line picture for Equation (2). In the upper-half plane, $\frac{dy}{dt} < 0$ so all vectors of the vector field point downward. Above an equilibrium point, the vector field points directly down. Above a point on the phase line with the arrow pointing right, the vector field points down and to the right. Above a point on the phase line with the arrow pointing left, the vector field points down and to the left. In the lower-half plane, the vector field points upward and a similar analysis can be applied.

For fixed r, to draw the phase portrait of Equation (3), first draw the phase line of (2), draw the trajectories directly above and directly below the equilibrium points, and then fill in a representative set of trajectories.

2.1   Draw the phase portrait of the system in Equation (3) with $r = 1$.

2.2   Draw the phase portrait of the system in Equation (3) with $r = -1$.

2.3   Why is the bifurcation in Equation (2) called a saddle-node bifurcation?

Another family exhibiting a saddle-node bifurcation when the parameter r is 0 is

$$\frac{dx}{dt} = y$$

$$\frac{dy}{dt} = x^2 - y - r$$

$$(4)$$

Open the **2-D Saddle Node Bifurcation** tool to see an animation of how the phase portrait of Equation (4) changes as r varies through 0. Equation (4) does not define a product family, but it is essentially one-dimensional in the sense that its phase portraits for various values of r are the same as the phase portraits for Equation (3) up to "continuous change of coordinates."

2.4   Find all equilibria for Equation (4) when $r = -1$ and when $r = 1$, linearize Equation (4) at the equilibria, and verify how it behaves near the equilibria.

2.5   For the family of first-order differential equations $\dot{x} = r^2 + x^2 - 1$, draw the bifurcation diagram. Form the product family. For a representative set of values of r, fix r and draw the phase portrait of the system. Describe what happens at each bifurcation of the product family.

Note: The Hopf bifurcation and product bifurcations are just two illustrations of the kinds of bifurcations planar autonomous systems can undergo.

Lab 25: Tool Instructions

Chemical Oscillator Tool

Setting Initial Conditions

Click the mouse on the graphing plane to set the initial conditions for a trajectory.

Clicking in the plane while the trajectory is being drawn will start a new trajectory.

Parameter Sliders

Use the sliders to set the parameter b.

Press the mouse down on the slider knob and drag the mouse back and forth, or click the mouse in the slider channel at the desired value for the parameter.

Buttons

Click the mouse on the [Clear] button to remove all the trajectories from the graph.

Click the mouse on the [Pause] button to stop a trajectory without canceling it.

Click the mouse on the [Continue] button to resume the motion of the paused trajectory.

2-D Saddle-Node Bifurcation Tool

Buttons

Click on the arrow buttons to control the animation sequence of the path of the parameter k. Use the double arrow buttons to play the sequence forward and backward. Use the single arrow buttons to advance and reverse the sequence one frame at a time.

Click on the single arrow button to stop the sequence.

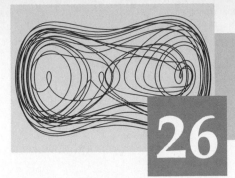

26

Chaos in Forced Nonlinear Oscillators

What happens to the behavior of solutions as $t \to \infty$ in non autonomous, nonlinear, planar differential equations that model some simple applications?

1. Forced Damped Nonlinear Pendulum

The differential equation for the forced, damped, nonlinear pendulum is

$$\ddot{\theta} = -\sin\theta - b\dot{\theta} + A\cos\omega t.$$

Here A is the magnitude of the forcing function, ω is the frequency of the forcing function, and b measures damping. Letting $x = \theta$ and $y = \dot{\theta}$, we can write this as a system of two first order equations:

$$\frac{dx}{dt} = y$$

$$\frac{dy}{dt} = -\sin x - by + A\cos\omega t.$$

Open the **Forced Damped Pendulum** tool.

1.1 Why isn't there a vector field drawn in the phase plane?

1.2 What difference is there in the solution curves of a planar autonomous system and a planar nonautonomous system?

1.3 With the default parameters ($A = 1.50$, $b = .50$, $\omega = 0.67$), how do the pendulum and corresponding solution curves behave as $t \to \infty$?

1.4   Change A to 1.35 and observe that the solution curves approach periodic behavior. Sketch the periodic orbit.

1.5   Change A to 1.45. What happens to the periodic orbit?

1.6   Change A to 1.47. What happens to the periodic orbit?

Systems that exhibit chaotic behavior often have values of the parameters where the solutions exhibit periodic behavior, as in Exercise **1.4**. Often as one changes the period slightly it will double, as in Exercise **1.5** and Exercise **1.6**, leading to the chaotic behavior of Exercise **1.3**.

1.7   If you wanted to make the origin an attractor, what strategy might you use for the parameters?

1.8   Do you think the sort of chaotic behavior of the solution curves as $t \to +\infty$ exhibited by the forced nonlinear pendulum could occur in a planar autonomous system? What are the limitations on the behavior a solution curve can exhibit as $t \to +\infty$ in a planar autonomous system?

2. Poincaré Section

The orbits of a nonautonomous system of two equations are a mess on a phase plane, because solutions can cross. It takes a three-dimensional phase space with axes for—typically—position, velocity, and time to sort solutions out to an uncrossed state. When we look at a planar representation, we are seeing a three-dimensional trajectory projected onto a plane. Though solutions for most systems do not cross in phase space, they often appear to cross when projected onto a plane. A loop floating in phase space that looks like an ellipse from above might look like an infinity symbol from the front.

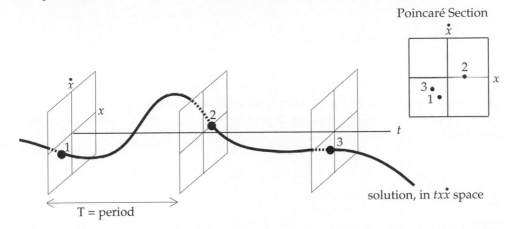

The orbits in the phase portrait consist of all points on a solution curve. To construct a Poincaré section, we do not plot every point on an orbit, only points on planes, parallel to the xx'-plane at evenly spaced time intervals in the phase space. In the case of the **Forced Damped Pendulum: Poincare Section** and **Forced Damped Pendulum: Nine Sections** tools, the time interval is $T = 2\pi/\omega$. Starting with the point (x_1, y_1) in the plane $t = 0$, the trajectory intersects the plane $t = T$ at the point (x_2, y_2) and the plane $t = 2T$ at the point (x_3, y_3). We generate a sequence of points $(x_1, y_1), (x_2, y_2), (x_3, y_3), \ldots$. These are the points plotted in the Poincaré section (they are labeled 1,2,3,…in the figure). Use the **Forced Damped Pendulum: Poincare Section** tool to get an idea of the concept of the Poincaré section.

The trajectory meeting the plane $t = T$ at (x_2, y_2, T) meets the plane $t = 2T$ at the point $(x_3, y_3, 2T)$. Because the forcing function is periodic with period T, the trajectory starting at $(x_2, y_2, 0)$ meets the plane $t = T$ at the point (x_3, y_3, T), and the trajectory from $(x_2, y_2, 0)$ to (x_3, y_3, T) is identical to the trajectory from (x_2, y_2, T) to $(x_3, y_3, 2T)$ translated left by T. Thus, in describing the dynamics of the forced pendulum, we can restrict our attention to the part of xyt-space between the planes $t = 0$ and $t = T$ — when a trajectory leaves this part of space through the plane $t = T$ at the point (x, y, T), it is considered to reenter this part of space through the plane $t = 0$ at the point $(x, y, 0)$. The Poincaré section is a slice through the figure consisting of the set of trajectories generated this way.

2.1   If a trajectory is periodic with period $3T$, how will it meet the part of space between the planes $t = 0$ and $t = T$?

2.2   How does the trajectory in Exercise **1.3** meet the part of space between the planes $t = 0$ and $t = T$?

The **Forced Damped Pendulum: Nine Sections** tool displays nine Poincaré sections taken at $\pi/(4\omega)$ time intervals in phase space. Use this tool to explore the next two questions (as usual, the **[Clear Transients]** button is useful in showing the attractor).

2.3   Will the Poincaré section depend on the sampling time? Discuss.

2.4   Why is the first picture the same as the last picture?

2.5   What should the Poincaré section look like if there is an attracting periodic orbit? *Hint:* Experiment with the parameter values that gave periodic orbits above. Use the **Forced Damped Pendulum: Poincare Section** tool.

3. Duffing's Equation

Consider a mechanical system in which a flexible steel beam is attached to the top, center of a box. There are magnets on either side of the box to which the beam is equally attracted. Duffing's equation describes the motion of the tip of the beam as the box is moved from side to side. The differential equation for the Duffing oscillator is:

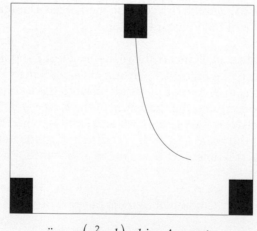

$$\ddot{x} = -x\left(x^2 - k\right) - b\dot{x} + A\cos\omega t.$$

Here, A is the magnitude of the forcing function, ω is the frequency of the forcing function, b is the damping, and k is related to the stiffness of the beam and the strength of the magnets. The center of the box is at $x = 0$. Letting $y = \dot{x}$ we can write this equation as a system of two first-order equations:

$$\frac{dx}{dt} = y$$

$$\frac{dy}{dt} = -x\left(x^2 - k\right) - by + A\cos\omega t.$$

Open the **Duffing Oscillator** tool.

3.1   With the default parameters ($k = 1$, $b = .15$, $A = .30$, $\omega = 1.00$), how do the tip of the steel beam and the corresponding solution curves behave as $t \rightarrow +\infty$? Note that there are two completely different types of behavior depending on the choice of initial conditions. Describe both.

3.2   Take parameters $k = 1$, $b = .20$, $A = .30$, and $\omega = .60$ and sketch the resulting periodic orbit.

3.3   If you change b to .15, how does the periodic orbit change?

3.4   If you change b to .14, how does the periodic orbit change?

You are invited to explore other parameter values for the **Forced Damped Pendulum** and **Duffing Oscillator** tools.

Lab 26: Tool Instructions

Forced Damped Pendulum Tool

Setting Initial Conditions

Click the mouse on the graphing plane to set the initial conditions for a trajectory.

Clicking in the plane while a trajectory is being drawn will start a new trajectory.

Parameter Sliders

Use the sliders to set the damping constant b, the forcing amplitude A, and the forcing frequency w. Press the mouse down on the slider knob for the parameter you want to change and drag the mouse back and forth, or click the mouse in the slider channel at the desired value for the parameter.

Time Series Buttons

The buttons labeled

[ ] **position**
[ ] **velocity**
[ ] **acceleration**

toggle the time series on and off.

Other Buttons

Click the mouse on the **[Clear]** button to remove all the trajectories from the graphs.

Click the mouse on the **[Clear Transients]** button to remove all transient data from the graph without disrupting the active trajectory.

Click the mouse on the **[Pause]** button to stop a trajectory without canceling it.

Click the mouse on the **[Continue]** button to resume the motion of the paused trajectory

Force Damped Pendulum: Poincare Section Tool

Setting Initial Conditions

Click the mouse on the graphing plane to set the initial conditions for a trajectory.

Clicking in the plane while a trajectory is being drawn will start a new trajectory.

Parameter Sliders

Use the sliders to set the damping constant b, the forcing amplitude A, and the forcing frequency w. Press the mouse down on the slider knob for the parameter you want to change and drag the mouse back and forth, or click the mouse in the slider channel at the desired value for the parameter.

Time Series Buttons

The buttons labeled

[ ] **position**
[ ] **velocity**

toggle the time series on and off.

Other Buttons

Click the mouse on the **[Clear]** button to remove all the trajectories from the graphs.

Click the mouse on the **[Clear Transients]** button to remove all transient data from the graph without disrupting the active trajectory.

Click the mouse on the **[Pause]** button to stop a trajectory without canceling it.

Click the mouse on the **[Continue]** button to resume the motion of the paused trajectory

Forced Damped Pendulum: Nine Sections Tool

Setting Initial Conditions

Click the mouse on any of the nine graphing planes to set the initial conditions for a trajectory.
Clicking in the plane while a trajectory is being drawn will start a new trajectory.

Parameter Sliders

Use the sliders to set the damping constant b, the forcing amplitude A, and the forcing frequency w.
Press the mouse down on the slider knob for the parameter you want to change and drag the mouse
back and forth, or click the mouse in the slider channel at the desired value for the parameter.

Other Buttons

Click the mouse on the [Clear] button to remove all the trajectories from the graphs.
Click the mouse on the [Clear Transients] button to remove all transient data from the graph without disrupting the active trajectory.

Duffing Oscillator Tool

Setting Initial Conditions

Click the mouse on the graphing plane to set the initial conditions for a trajectory.
Clicking in the plane while a trajectory is being drawn will start a new trajectory.

Parameter Sliders

Use the sliders to set the damping constant b, the forcing amplitude A, and the forcing frequency w,
and the ratio of mechanical elastic force and electromagnetic force k.

Press the mouse down on the slider knob for the parameter you want to change and drag the mouse
back and forth, or click the mouse in the slider channel at the desired value for the parameter.

Time Series Buttons

The buttons labeled
   [ ] **position**
   [ ] **velocity**
   [ ] **acceleration**
toggle the time series on and off.

Other Buttons

Click the mouse on the [Clear] button to remove all the trajectories from the graphs.
Click the mouse on the [Clear Transients] button to remove all transient data from the graph with-
out disrupting the active trajectory.
Click the mouse on the [Pause] button to stop a trajectory without canceling it.
Click the mouse on the [Continue] button to resume the motion of the paused trajectory

27 The Lorenz Equations

In 1963, Edward Lorenz published a paper on a simplified model for convection that has become a cornerstone for the field of nonlinear dynamics. We will investigate changes in the behavior of solutions of the Lorenz equation as the parameter r is varied.

1. The Lorenz Equations

The Lorenz system includes three equations and three parameters. For this lab, one of the parameters is taken to be 10 and another to be 8/3. The remaining parameter r is called the **Rayleigh number**. It is proportional to the difference in temperature from the warm base of a convection cell to the cooler top. The variable $x(t)$ is the speed of rotation of the cell. The other two variables $y(t)$, $z(t)$ represent temperature distribution in the cell.

$$\dot{x} = 10(y - x)$$
$$\dot{y} = rx - y - xz \qquad (1)$$
$$\dot{z} = xy - \frac{8}{3}z$$

In some of the Lorenz tools, the convection cell is represented by a rectangle with a red (hot) base and a blue (cool) top. The variable $x(t)$ is represented by a square rotating inside the rectangle. If the square is not rotating, then $x(t)$ is zero.

2. Sensitive Dependence on Initial Conditions

The **Lorenz Equations: Discovery 1963** illustration recreates what Lorenz described at the time. His computer calculated a trajectory using six places of accuracy, but printed the list of values using only three. When he stopped the trajectory, he used the three-place data at the stopping time to start it up again and was puzzled when the results did not match a previous run of the trajectory that did not have a stop in the middle. That is when he concluded that the missing last three places of accuracy must be important.

Lorenz observed that nonlinear systems exhibit **sensitive dependence on initial conditions**—very small changes in initial conditions can make very large differences in long-term behavior. Open the **Lorenz**

Equations: Sensitive Dependence tool for an illustration of sensitive dependence on initial conditions for the trajectory $(x(t), y(t), z(t))$ when $r = 28$. Note also that any trajectory, when $r = 28$, approaches a **strange attractor**, that is, an attractor that is not an equilibrium, nor a cycle, nor a finite graph.

3. Equilibria of the Lorenz System

The origin is an equilibrium of the Lorenz equation.

3.1   Linearize the equations in Equation (1) at the origin. Find the eigenvalues of the linearized system and show that the origin is an attractor for the linearized system if $r < 1$, but is unstable if $r > 1$.

3.2   Show, in fact, that for $r < 1$, every trajectory is asymptotic to the origin by showing that

$$E(x, y, z) = \frac{x^2}{10} + y^2 + z^2 \text{ is a Lyapunov function for the Lorenz equation. That is, show that}$$

$$\frac{d}{dt} E(x(t), y(t), z(t)) < 0 \text{ if } r < 1 \text{ and } (x, y, z) \neq (0, 0, 0).$$

3.3   Solve the Lorenz equations with initial conditions $(x(0), y(0), z(0)) = (0, 0, z_0)$ and show that for every value of r, every point on the z-axis is attracted to the origin.

3.4   Show that the Lorenz equations are symmetric with respect to the z-axis.

For $r > 1$, the Lorenz equation has two additional equilibria, C^+ and C^-. These two points lie at the center of each of the two "butterfly wings" comprising the strange attractor when $r = 28$.

3.5   Find formulas for C^+ and C^-.

3.6   Linearize Equation (1) at C^+.

The characteristic equation for the linear equation in Exercise **3.6** is:

$$\lambda^3 + \frac{41}{3}\lambda^2 + \left(\frac{8}{3}r + \frac{80}{3}\right)\lambda + \frac{160}{3}(r-1) = 0$$

For $r > 0$, this equation has solutions $\lambda_0 < 0$ *and* $a \pm bi$. Thus, if $a < 0$, C^+ is an attractor and if $a > 0$, C^+ is a repellor. The point C^+ changes from attractor to repellor at a value of r for which $a = 0$. The characteristic equation then has the form:

$$(\lambda - \lambda_0)(\lambda - bi)(\lambda + bi) = (\lambda - \lambda_0)(\lambda^2 + b^2) = \lambda^3 - \lambda_0\lambda^2 + b^2\lambda - \lambda_0 b^2 = 0$$

3.7   By comparing coefficients of the two forms of the characteristic equation, we get three equations in the three unknowns λ_0, b, r. Solve for r to get the value of r at which C^+ and C^- turn from attractor to repellor.

We have seen that the origin is an attractor for $r < 1$. As r increases, a bifurcation occurs at $r = 1$. For $r > 1$, the origin becomes an unstable equilibrium and two attracting equilibria, C^+ and C^-, are born. At $r = 24.74$, C^+ and C^- become unstable. The origin, C^+ and C^- are the only equilibria. Thus for $r > 24.74$ there are no attracting equilibria. However, chaotic behavior already begins at $r = 24.06$.

For a fixed value of the parameter r, there is a large ellipsoidal trapping region for the Lorenz equation such that all trajectories enter that region and none leave. See, for example, Exercise **9.22**, p. 343 of *Nonlinear Dynamics and Chaos*, by Steven H. Strogatz. Thus within that region, there must be an attractor: a point, a cycle, or a strange attractor.

Use the **Lorenz Equations: Parameter Grid** $0 \le r \le 30$ tool and the **Lorenz Equations: Phase Plane** $0 \le r < 30$ tool to verify the asymptotic behavior of solution curves for various values of r we have described up to this point.

4. Other Types of Behavior

Nonlinear systems, which exhibit chaotic behavior for certain choices of parameters, exhibit a variety of interesting behaviors. Open the **Lorenz Equations: Phase Plane $0 \leq r \leq 320$** tool to explore the ideas in this section.

Transient Chaos: Chaotic behavior begins when $r = 24.06$. When $r \leq 24.06$, all trajectories approach an equilibrium. For values of r less than 24.06, but close to 24.06 and for some choices of initial conditions, trajectories will appear to be chaotic for an interval of time before approaching the equilibrium. This phenomenon is called **transient chaos**.

Periodic Windows: Even though for many choices of parameter $r > 24.74$, the behavior of the system is chaotic, there are intervals of parameters on which the trajectories approach one or more attracting cycles. Such parameter intervals are called **periodic windows**. Periodic windows occur around $r = 220, 166, 133, 126.5, 114, 100$.

We can code trajectories with strings of 0's and 1's, putting a 0 when the trajectory loops around C^+ and a 1 when the trajectory loops around C^-. If the trajectory approaches an attracting cycle, the string of 0's and 1's will eventually begin to repeat. We can code an attracting cycle with the finite string of 0's and 1's required until the trajectory repeats itself. If such a string encodes an attracting cycle, then any cyclic permutation of it will also encode the attracting cycle.

4.1   Find an attracting cycle when $r = 100.2$ and give the finite string of 0's and 1's that codes this cycle.

4.2   Use the fact that the Lorenz equations are symmetric about the z-axis to find another attracting cycle when $r = 100.2$ and give the finite string of 0's and 1's that codes this cycle.

4.3   Decrease r to 99.80 and find an attracting cycle. Give the finite string of 0's and 1's that codes this cycle. How does this string compare to one of the strings in Exercise **4.1** or **4.2**?

Period Doubling, Noisy Periodicity: The cycle found in Exercise **4.3** is said to have twice the period of the cycles in Exercise **4.1** and **4.2**. Taking slightly smaller values of the parameters would produce cycles with 4, 8, 16,... times the period of the original cycles. This phenomenon is called **period doubling** and occurs as one approaches the lower end of the other periodic windows. The software is not sharp enough to detect all this, but does clearly show cycles with four times the original period in the periodic windows around $r = 220$ and $r = 160$. If we continue past parameter values that produce period doubling, we encounter **noisy periodicity**, that is the trajectories are no longer cycles but they still have the same patterns of 0's and 1's as the cycles in the periodic windows. Finally, if we reduce the parameter still further, we encounter total chaos.

Intermittency: In the previous paragraph we explored how the attractors change near the lower end of a periodic window. In this paragraph we explore how the attractors behave near the upper end of a periodic window. Let $r = 166.06$ and find an attracting cycle. Now let $r = 166.07$. The solutions approach an attractor that for some period of time behaves as if it is a cycle and then behaves chaotically for a while before settling down into periodic behavior before once again displaying a burst of chaotic behavior—a very strange attractor indeed. This phenomenon is called **intermittency** and can also be observed above the upper end of the other periodic windows.

The Lorenz Map

Let $z_0, z_1, z_2,\ldots$ be the successive maximum values of a solution $z(t)$ for some choice of initial condition. The **Lorenz Equations: Zmax Map** tool plots the points (z_{n-1}, z_n) for $n = 1,2,3,\ldots$. Lorenz used this remarkable plot (called the **Lorenz map**) to give a heuristic argument that the apparent chaotic behavior of the Lorenz equations for certain values of r is real—that is, there are no attracting cycles for these values of r.

Note: This lab has attempted to give a first idea of how the behavior of the Lorenz equations changes with r. For further information on the Lorenz equation, see Steven H. Strogatz, *Nonlinear Dynamics and Chaos*, Addison-Wesley, Reading, MA: 1994, Chapter 9.

Lab 27: Tool Instructions

Lorenz Equations: Discovery 1963 Tool

Setting Initial Conditions

The initial conditions stated in the 1963 Lorenz paper are used in the first display, in combination with the parameter settings and numerical methods from the same source.

Click the mouse on the graphing plane to set the initial x value for a time series.

Lorenz Equations: Sensitive Dependence Tool

Setting Initial Conditions

Click the mouse on the graphing plane to set the initial conditions for a trajectory.

Clicking in the plane while a trajectory is being drawn will start a new trajectory.

Parameter Sliders

Use the sliders to set the Rayleigh constant r.

Press the mouse down on the slider knob and drag the mouse back and forth, or click the mouse in the slider channel at the desired value for the parameter.

Buttons

Click the mouse on the [Clear] button to remove all the trajectories from the graphs.

Click the mouse on the [Clear Transients] button to remove all transient data from the graph without disrupting the active trajectory.

Click the mouse on the [Pause] button to stop a trajectory without canceling it.

Click the mouse on the [Continue] button to resume the motion of the paused trajectory.

Click the mouse on the [Time Series] button to plot time series data from two slightly different values of the x variable.

Click the mouse on the [$|\Delta|$ Divergence] button to plot the divergence of two trajectories against time.

Click the mouse on the [XY] button to view the xy phase plane.

Click the mouse on the [XZ] button to view the xz phase plane.

Click the mouse on the [YZ] button to view the yz phase plane.

Lorenz Equations: Parameter Grid $0 < r \le 30$ Tool

Setting Initial Conditions

Click the mouse on any of the nine graphing planes to set the initial conditions for a trajectory.

Clicking in the plane while a trajectory is being drawn will start a new trajectory.

Lorenz Equations: Phase Plane $0 \le r \le 30$ Tool

Setting Initial Conditions

Click the mouse on the graphing plane to set the initial conditions for a trajectory.

Clicking in the plane while a trajectory is being drawn will start a new trajectory.

Parameter Sliders

Use the sliders to set the Rayleigh constant r.

Press the mouse down on the slider knob and drag the mouse back and forth, or click the mouse in the slider channel at the desired value for the parameter.

Buttons

Click the mouse on the **[Clear]** button to remove all the trajectories from the graphs.

Click the mouse on the **[Clear Transients]** button to remove all transient data from the graph without disrupting the active trajectory.

Click the mouse on the **[Pause]** button to stop a trajectory without canceling it.

Click the mouse on the **[Continue]** button to resume the motion of the paused trajectory.

Click the mouse on the **[XY]** button to view the xy phase plane.

Click the mouse on the **[XZ]** button to view the xz phase plane.

Click the mouse on the **[YZ]** button to view the yz phase plane.

Lorenz Equations: Phase Plane $0 \leq r \leq 320$ Tool

Setting Initial Conditions

Click the mouse on the graphing plane to set the initial conditions for a trajectory.

Clicking in the plane while a trajectory is being drawn will start a new trajectory.

Parameter Sliders

Use the sliders to set the Rayleigh constant r, or (1/100ths).

Press the mouse down on the slider knob and drag the mouse up and down, or click the mouse in the slider channel at the desired value for the parameter.

Buttons

Click the mouse on the **[Clear]** button to remove all the trajectories from the graphs.

Click the mouse on the **[Clear Transients]** button to remove all transient data from the graph without disrupting the active trajectory.

Click the mouse on the **[Pause]** button to stop a trajectory without canceling it.

Click the mouse on the **[Continue]** button to resume the motion of the paused trajectory.

Click the mouse on the **[XY]** button to view the xy phase plane.

Click the mouse on the **[XZ]** button to view the xz phase plane.

Click the mouse on the **[YZ]** button to view the yz phase plane.

Lorenz Equations: Zmax Map Tool

Setting Initial Conditions

Click the mouse on the graphing plane to set the initial conditions for a trajectory.

Clicking in the plane while a trajectory is being drawn will start a new trajectory.

Parameter Sliders

Use the slider to change the value for parameter r.

Press the mouse down on the slider knob and drag the mouse back and forth, or click the mouse in the slider channel at the desired value for the parameter.

Buttons

Click the mouse on the **[Clear]** button to remove all the trajectories from the graphs.

VI

Series Solutions and Boundary Value Problems

28 Maclaurin Series, Airy's Series

How are functions approximated by their Maclaurin polynomials?
How do the new functions that arise as solutions of the simple
differential equation $x'' + tx = 0$ behave and how are they similar to
familiar functions?

1. Power Series for Elementary Functions

Recall that a Maclaurin series is a Taylor series about $t_0 = 0$. The nth **Maclaurin polynomial** $p_n(t)$ (the sum of the terms of the Maclaurin series up to and including those of degree n) of a function $f(t)$ is the polynomial of degree n that, in some sense, best approximates the function at $t_0 = 0$.

$$p_n(t) = \sum_{k=0}^{n} f^{(k)}(0)\frac{t^k}{k!} \qquad (1)$$

The accuracy of the function can be judged by comparing it to the original function; the difference is the error. The error after n terms $E_n(t) = |f(t) - p_n(t)|$ is of **degree** $m > n$ in the sense that the error behaves like ct^m for t near 0. The nth Maclaurin polynomial is the unique polynomial of degree n that approximates $f(t)$ with error of degree $m > n$. In fact, sufficiently near 0, the error is very well approximated by the absolute value of the first deleted term of the series.

Use the **Maclaurin Series: Sine**, the **Maclaurin Series: Cosine**, the **Maclaurin Series: e^t** and the **Maclaurin Series:** $1/\sqrt{1+t}$ tools to investigate how these series converge to the function. Select different numbers of terms and see how close the series approximations are. Investigate the fact that, near 0, the error is well approximated by the absolute value of the first deleted term. Try different values of m to see how individual terms of the series, shown in orange, approximate the error shown on the lower graph.

235

1.1   For the following functions $f(t)$ and integers n, give the degree m of the error of the approximation of $f(t)$ by the nth Maclaurin polynomial:

    a. $f(t) = \sin(t)$ $n = 3;$ $m =$

    b. $f(t) = \cos(t)$ $n = 2;$ $m =$

    c. $f(t) = e^t$ $n = 2;$ $m =$

    d. $f(t) = f(t) = 1/\sqrt{1+t}$ $n = 2;$ $m =$

1.2   In the **Maclaurin Series: Sine** and **Maclaurin Series: Cosine** tools, the absolute value of the first deleted term gives a pretty good approximation of the error in the range selected. How good do you think the approximation would be further from 0? Why?

1.3   In the **Maclaurin Series: e^t** tool, the absolute value of the first deleted term does not appear to give as good an approximation to the error as in the **Maclaurin Series: Sine** and **Maclaurin Series: Cosine** tools. Why is this?

1.4   In the Maclaurin series for $1/\sqrt{1+t}$, the absolute value of the first deleted term does not appear to give as good an approximation to the error as for the other series. Why is this?

1.5   Show that for $f(t) = e^t$ and $n = 2$, the error is well approximated near zero by the absolute value of the first deleted term; that is, show that $\left| e^t - \left(1 + t + t^2/2! \right) \right|$ is well approximated by $|t|^3/3!$ near zero.

2. Additional Exercises

2.1   Use infinite series techniques to show that $x(t) = \sin(t)$ is a solution of the initial-value problem
$$x'' + x = 0, \ x(0) = 0, \ x'(0) = 1$$

2.2   Use infinite series techniques to show that $x(t) = \cos(t)$ is a solution of the initial-value problem
$$x'' + x = 0, \ x(0) = 1, \ x'(0) = 0.$$

3. Airy's Equation

Sir George Biddell Airy (1801–1892) was Astronomer Royal of England for many years and thus was interested in optics. Airy's equation, $x'' + tx = 0$ is used in studying the diffraction of light.

3.1   Open the **Airy's Equation** tool to see what the solutions look like. Sketch a phase plane solution in the xx'-plane.

3.2   Use the **Airy's Series: Cosine** tool to explore how the Maclaurin polynomials for the solution of Airy's equation $x'' + tx = 0$, with initial conditions $x(0) = 1$, $x'(0) = 0$, approximate the solution. What is the degree m of the error in the approximation of the Airy cosine function by the third Maclaurin polynomial? Why does the absolute value of the first deleted term appear to give so good an approximation to the error?

3.3   Use the **Airy's Series: Sine** tool to explore how the Maclaurin polynomials for the solution of $x'' + tx = 0$, with initial conditions $x(0) = 0$, $x'(0) = 1$, approximate the solution. What is the degree m of the error in the approximation of the Airy sine function by the first Maclaurin polynomial?

The differential equation $x'' + x = 0$ models a spring (without damping) with the spring constant equal to 1. Airy's equation $x'' + tx = 0$ can be thought of as modeling a spring with a variable spring "constant" t; that is, an aging spring that gets stiffer with time. Use the **Airy's Equation** tool to answer the following questions.

3.4   Compare and contrast the solutions $x(t)$ of these two equations satisfying the initial condition $x(0) = 1$, $x'(0) = 0$.

3.5   Compare and contrast the solutions $x(t)$ of these two equations satisfying the initial condition $x(0) = 0$, $x'(0) = 1$.

3.6   With initial conditions $x(0) = 1$, $x'(0) = 0$, the integral curve in the phase plane for $x'' + x = 0$ is a circle and for Airy's equation is something different. For $t > 0$, describe the difference and explain what differences in the graph of the corresponding solution curves $x(t)$ help explain that difference.

3.7   In general, at what points of the tx-plane will a solution of Airy's equation have a point of inflection?

3.8   Describe the behavior of the general solution $x(t)$ of Airy's equation when $t > 0$.

3.9   Describe the behavior of the general solution $x(t)$ of Airy's equation when $t < 0$.

3.10   The graph of the Airy cosine has a cusp in the phase plane. Let the lower graph finish plotting to see the cusp appear. Does the cusp indicate a lack of differentiability on the part of the Airy Cosine?

4. Additional Exercises

4.1   Use infinite series techniques to give two independent Maclaurin series solutions $x_1(t)$, $x_2(t)$ of the Airy's equation satisfying the initial conditions $x_1(0) = 1$, $x'_1(0) = 0$ and $x_2(0) = 0$, $x'_2(0) = 1$.

4.2   Use the substitution $u(t) = x'(t)/x(t)$ to transform the Airy equation into the equation $u'(t) = -u^2(t) - t$.

4.3   Some books give Airy's equation in the form $x'' - tx = 0$. Show that this equation is obtained from the equation $x'' + tx = 0$ by the change of coordinates $(x,t) \rightarrow (-x,-t)$. What effect does this change of coordinates have on solutions in the tx-plane?

Note: For a detailed analysis of the solutions to Airy's equation, see Hubbard, McDill, Noonburg, and West, *College Mathematics Journal* 25 (1994): 419–43.

Lab 28: Tool Instructions

Maclaurin Series: Sine Tool

Partial Series Buttons
Click a term button to select the number of terms for the partial sum.
Click the [All] button to overlay the first five approximations.

Parameter Slider
Use the slider to set the value of m in the error term.
Press the mouse down on the slider knob and drag the mouse back and forth, or click the mouse in the slider channel at the desired value for the parameter.

Maclaurin Series: Cosine Tool

Partial Series Buttons
Click a term button to select the number of terms for the partial sum.
Click the [All] button to overlay the first five approximations.

Parameter Slider
Use the slider to set the value of m in error term.
Press the mouse down on the slider knob and drag the mouse back and forth, or click the mouse in the slider channel at the desired value for the parameter.

Maclaurin Series: e^t Tool

Partial Series Buttons
Click a term button to select the number of terms for the partial sum.
Click the [All] button to overlay the first five approximations.

Parameter Slider
Use the slider to set the value of m in the error term.
Press the mouse down on the slider knob and drag the mouse back and forth, or click the mouse in the slider channel at the desired value for the parameter.

Maclaurin Series: $1/\sqrt{1+t}$ Tool

Partial Series Buttons
Click a term button to select the number of terms for the partial sum.
Click the [All] button to overlay the first five approximations.

Parameter Slider
Use the slider to set the value of m in the error term.
Press the mouse down on the slider knob and drag the mouse back and forth, or click the mouse in the slider channel at the desired value for the parameter.

Airy's Equation Tool

Setting Initial Conditions
Click the mouse on the graphing plane to set the initial conditions for a trajectory or a point for a vector.
Clicking in the plane while a trajectory is being drawn will start a new trajectory.

Buttons
Click the mouse on the **[Cosine]** button to set the initial conditions for Airy's Cosine: $x(0) = 1$, $v(0) = 0$.
Click the mouse on the **[Sine]** button to set the initial conditions for Airy's Sine: $x(0) = 0$, $v(0) = 1$.
Click the mouse on the **[Clear]** button to remove all trajectories from the graph.

Airy's Series: Cosine Tool

Partial Series Buttons
Click a term button to select the number of terms for the partial sum.
Click the **[All]** button to overlay the first five approximations.

Parameter Slider
Use the slider to set the value of m in the error term.
Press the mouse down on the slider knob and drag the mouse back and forth, or click the mouse in the slider channel at the desired value for the parameter.

Airy's Series: Sine Tool

Partial Series Buttons
Click a term button to select the number of terms for the partial sum.
Click the **[All]** button to overlay the first five approximations.

Parameter Slider
Use the slider to set the value of m in the error term.
Press the mouse down on the slider knob and drag the mouse back and forth, or click the mouse in the slider channel at the desired value for the parameter.

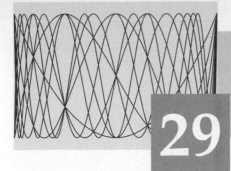

29 Special Functions (from Series Solutions)

Tools Used in Lab 29

Chebyshev's Equation
Lissajous Figures
Bessel Function: 1st Kind

Power series solutions of differential equations give rise to new functions, functions that cannot be expressed in terms of sines, cosines, exponentials, logarithms, roots, or any familiar functions. How do these new functions behave and what are their properties?

1. Chebyshev's Equation

For $n = 0, 1, 2, \ldots$ Chebyshev's equation, $(1-t^2)x'' - tx' + n^2x = 0$, $-1 < t < 1$, has polynomial solutions. These polynomials are multiples of the **Chebyshev polynomials** $T_n(t) = \cos(n \arccos(t))$. The nth Chebyshev polynomial $T_n(t)$ can be obtained by expressing $\cos n\theta$ in terms of $\cos \theta$ and substituting $t = \cos \theta$.

The formula $T_n(t) = \cos(n \arccos(t))$ makes sense even when n is not an integer and in fact, for any real number n, $T_n(t)$ is a solution of Chebyshev's equation. The function $S_n(t) = \sin(n \arccos(t))$ is also a solution of the Chebyshev equation, which is independent of $T_n(t)$ if $n > 0$.

Use the **Chebyshev's Equation** tool to answer the following questions.

1.1 Examine the graphs of the first few Chebyshev polynomials and give some properties of $T_n(t)$.

1.2 What is the value of: $T_n(0)$? $T'_n(0)$? $T_n(-1)$?

1.3   Give the formula for $T_{1/2}(t)$. What standard trig identity is this formula a restatement of? *(Hint: Let $t = \cos(\theta)$.)*

1.4   Give a formula for $T_2(t)$. What standard trig identity is this formula a restatement of?

2. DeMoivre's Formula: $\cos n\theta + i \sin n\theta = (\cos \theta + i \sin \theta)^n$

2.1   Use DeMoivre's formula and the trig identity $\cos^2 \theta + \sin^2 \theta = 1$ to find formulas for the first few Chebyshev polynomials and verify that these are the polynomials given by the **Chebyshev's Equation** tool.

2.2   Compute the first few polynomials given by the recurrence relation, $T_0(t) = 1$, $T_1(t) = t$, $T_n(t) = 2tT_{n-1}(t) - T_{n-2}(t)$, $n \geq 2$, and verify that these are the polynomials given by the **Chebyshev's Equation** tool.

2.3   Use infinite series methods to give a recurrence relation for the coefficients of the Maclaurin series expansion of the solutions to Chebyshev's equation with n arbitrary. For the first few non-negative integers, verify that Chebyshev's equation has a polynomial solution that is a multiple of the Chebyshev polynomial.

2.4 By differentiating $T_n(t) = \cos(n \arccos(t))$ twice, show that $T_n(t)$ is a solution of Chebyshev's equation for all real numbers n. Notice that your computations just as easily show that $x(t) = R \cos(\arccos(t) - \delta)$ is a solution for R and δ arbitrary.

2.5 Use DeMoivre's formula and the trig identity $\cos^2\theta + \sin^2\theta = 1$ to find formulas for the first few of the functions $S_n(t)$ and verify that these are the polynomials given by the **Chebyshev's Equation** tool.

2.6 Show that $T_n(t)$ and $S_n(t)$ are independent for $n > 0$.

2.7 Show that $c_1 \cos\theta + c_2 \sin\theta = R\cos(\theta - \delta)$ where $R = \sqrt{c_1^2 + c_2^2}$ and $\tan\delta = c_2 / c_1$ (thus $c_1 = R\cos\theta$ and $c_2 = R\sin\theta$). Conclude that the formula given on the **Chebyshev's Equation** tool is the correct formula for the general solution of Chebyshev's equation for $n > 0$.

2.8 When $n = 0$, $x(t)$ identically constant is a solution of Chebyshev's equation. Find a second independent solution of Chebyshev's equation when $n = 0$.

3. Lissajous Figures

Jules Antoine Lissajous (1822–1880) was interested in the physics of wave motion. Lissajous obtained his figures, in the context of acoustics, from the superposition of the vibrations of tuning forks, vibrating about mutually perpendicular axes. If Lissajous knew the characterisics (frequency and phase shift) of one tuning fork, he was able to deduce the characteristics of the other tuning fork from the **Lissajous figure**.

Lissajous figures are graphs in the xy-plane given parametrically by

$$x(t) = \cos(2m\pi t - a)$$
$$y(t) = \cos(2n\pi t - b)$$

We will see that Lissajous figures are related to Chebyshev polynomials.

For the following questions, use the **Lissajous Figures** tool.

3.1   If $a = 0$, $b = \pi/2$, $m = n = 1$, what is the Lissajous figure? Why?

3.2   If $m = n = 1$, how can one detect whether $x(t)$ and $y(t)$ are in phase, that is, that $a - b = 2k\pi$?

3.3   If $m = 1$ and $a = b = 0$, what is the Lissajous figure? Why?

3.4   How can we detect the ratio of the frequency of the wave $x(t)$ to the frequency of the wave $y(t)$, that is, the ratio m/n?

3.5   What will the Lissajous figure look like if m/n is not an integer?

4. Other Equations with Maclaurin Series Solutions

There are other interesting families of equations that have important applications, which have series solutions that are generally not expressible in terms of the elementary functions, but which have polynomial solutions for certain values of the parameter. Among such families of equations are

Hermite's Equation: $x'' - 2tx + 2nx = 0$;

Laguerre's Equation: $tx'' + (1 - t)x' + nx = 0$, $t > 0$;

Legendre's Equation: $(1 - t^2)x'' - 2tx' + n^2x = 0$, $-1 < x < 1$.

You can use a differential equation graphical solver to explore and analyze one of these.

5. Bessel's Equation

Bessel's equation of order p, $t^2x'' + tx' + (t^2 - p^2)x = 0$, $0 < t < \infty$, is singular at $t = 0$ and thus does not generally have Maclaurin series solutions. However, Frobenius's method gives a series solution, denoted $J_p(t)$, called the Bessel function of first kind of order p. The **Bessel Function: 1st Kind** tool graphs $J_p(t)$.

Lab 29: Tool Instructions

Chebyshev's Equation Tool

Parameter Slider

Use the slider to set the parameter n.
Chebyshev polynomials are found at integer values of n.
Press the mouse down on the slider knob and drag the mouse back and forth, or click the mouse in the slider channel at the desired value for the parameter.

Buttons

Click the [Chebyshev Function] button to get three parameters, r, n, s, and the buttons [Chebyshev Cosine] and [Chebyshev Sine].

Lissajous Figures Tool

Parameter Slider

Use the slider to change the values for the parameter a, m, b, and n.
Press the mouse down on the slider knob for the parameter you want to change and drag the mouse back and forth, or click the mouse in the slider channel at the desired value for the parameter.

Bessel's Function: 1st Kind Tool

Parameter Slider

Use the slider to set p, the order of the function.
Press the mouse down on the slider knob and drag the mouse back and forth, or click the mouse in the slider channel at the desired value for the parameter.

Boundary Value Problems

30

The vibration of a string, the motion of a drumhead are examples of physical systems whose behavior is determined by values of a function at the two endpoints of an interval rather than just at the initial point.

1. The Vibrating String

A mode of vibrating string satisfies the differential equation:

$$y'' + \lambda y = 0 \tag{1}$$

The ends of the string are held fixed so $y(0) = y(1) = 0$. Values of λ for which a boundary value problem has a non-trivial solution are called **eigenvalues.**

1.1 Solve Equation (1) and use the solution to figure out the eigenvalues.

Open the **Boundary Values: Eigenvalues** tool to verify that your eigenvalues in Exercise **1.1** are correct and to see some solutions of Equation (1). Be sure to click on the button $y(1) = 0$. The technique you use with the **Boundary Values: Eigenvalues** tool to look for eigenvalues is called **shooting**; you try a value of λ and "shoot" for the boundary condition $y(1) = 0$. You adjust your aim until you hit the boundary condition. In case you are a poor shot, the tool provides a "find closest eigenvalue" button.

Generally, if the string is plucked or strummed or somehow caused to vibrate, the equation of motion of the string will be given by

$$y(x,t) = \sum_{n=1}^{\infty} a_n \cos(n\pi ct - \delta_n)\sin n\pi x$$

The constant c depends on the string, and the factors $\sin n\pi x$ are the solutions to Exercise **1.1** satisfying the initial conditions $y(0) = y(1) = 0$, and are called the **fundamental modes** of vibration. Each one of the fundamental modes is multiplied by a term $a_n \cos(n\pi ct - \delta_n)$ to give the vibratory motion in that mode — a_n and δ_n are the amplitude and phase shift of the vibration in the nth fundamental mode. The total motion of the string is the sum of the motions in the fundamental modes. How each fundamental mode contributes is determined by a_n and δ_n, which in turn are determined by the initial conditions, which will be different depending on how the string is caused to vibrate.

2. A Different Set of Boundary Conditions

Instead of the boundary conditions $y(0) = y(1) = 0$, one could take the boundary conditions $y(0) = y'(1) = 0$ for Equation (1).

2.1   Use the solution to Equation (1) from Exercise **1.1** to figure out the eigenvalues with this new boundary condition.

Use the **Boundary Values: Eigenvalues** tool to verify that your eigenvalues in Exercise **2.1** are correct and to see some solutions of Equation (1) with these new boundary conditions. Be sure to click on the button $y'(1) = 0$.

3. Solvability and Uniqueness

In this section we consider the equation:

$$y'' + ky = \sin 2x, k > 0 \tag{2}$$

and look for solutions to the boundary value problem $y(0) = y(\pi) = 0$. Because of the $\sin 2x$ on the right hand side, the values k for which there is a solution are not called eigenvalues, which is why we use k instead of λ. Also, the k in Equation (2) behaves differently from the λ in Equation (1). Open the **Boundary Values: Solvability and Uniqueness** tool. This tool allows you to "shoot" for the boundary condition $y(\pi) = 0$ both by varying k and by varying $y'(0)$. For the solutions of Equation (1) satisfying $y(0) = 0$, varying $y'(0)$ did not change whether or not $y(x)$ was a solution of the boundary value problem. Here the situation is different.

3.1   Take $k = 4$. Can you find a solution for the boundary value problem? What value(s) can $y(\pi)$ assume?

3.2   Take $k = 5$. Can you find a solution of the boundary value problem? Is the solution unique? That is does more than one value of $y'(0)$ give a solution?

3.3   Take $k = 9$. Can you find a solution of the boundary value problem? Is the solution unique? That is does more than one value of $y'(0)$ give a solution?

3.4   Solve Equation (2) with initial conditions $y(0) = 0$.

3.5   Use your answer to Exercise **3.4** to tell for what values of k a solution to the boundary value problem of Equation (2) with $y(0) = y(\pi) = 0$ will exist and for what values of k the solution will be unique. When the solution exists (whether or not it is unique) indicate what value(s) of $y'(0)$ give a solution. If the solution does not exist, indicate why.

Note: Equation (2) replaced by Equation (3) (that is, with the 2 on the right side changed to an arbitrary positive number a) provides an interesting variation on Exercises **3.1–3.5**.

$$y'' + ky = \sin ax, k > 0 \qquad\qquad (3)$$

4. The Vibration of a Drumhead

Infinite series techniques can be used to solve Bessel's equation of order p:

$$t^2 x'' + tx' + \left(t^2 - p^2\right)x = 0, \quad 0 < t < \infty. \qquad\qquad (4)$$

Frobenius's method gives a solution, denoted $J_p(t)$, called the Bessel function of the first kind of order p. Bessel's equation is singular at $t = 0$ but for $p \geq 0$, $J_p(t)$ is continuous, $t \geq 0$. Open the **Bessel Function: 1st Kind** tool to recall the behavior of $J_p(t)$.

4.1   Use the **Bessel Function: 1st Kind** tool to find the first few zeros $\lambda_1, \lambda_2, \ldots$ of $J_0(t)$.

4.2   Show that under the change of independent variable $t = \lambda r$, Equation (4) is changed into the equation $R'' + \dfrac{1}{r}R' + \left(\lambda^2 - \dfrac{p^2}{r^2}\right)R = 0$. That is, $x(t)$ is a solution of $t^2 x'' + tx' + \left(t^2 - p^2\right)x = 0$ if and only if

$R(r) = x(\lambda r)$ is a solution of $R'' + \dfrac{1}{r}R' + \left(\lambda^2 - \dfrac{p^2}{r^2}\right)R = 0$.

Consider the boundary value problem (this is just the equation of Exercise **4.2** with $p = 0$):

$$R'' + \frac{1}{r}R' + \lambda^2 R = 0, R(0) = 1, R(1) = 0 \qquad\qquad (5)$$

4.3   Use the results of Exercises **4.1** and **4.2** to express the solutions of Equation (5) in terms of $J_0(t)$ and to find the first few eigenvalues of Equation (5) (values of λ for which Equation (5) has a solution).

Open the **Bessel Equation** tool to verify your answer in Exercise **4.3**. Again, you use the shooting method with this tool to find eigenvalues.

The motion of a circular drumhead of radius 1 struck exactly in the center is given by

$$H(t,r) = \sum_{k=1}^{\infty} a_k \cos(\lambda_k \omega t - \delta_k) J_0(\lambda_k r)$$

where ω is the natural frequency of the drum, the λ_k are the zeros of the Bessel function of the first kind of order zero, r is radial distance of a point from the center of the drumhead, a_k is amplitude, and δ_k is the phase shift of $J_0(\lambda_k r)$'s contribution to the vibration, and $H(t,r)$ is the height of a point at distance r from the center of the drumhead above the rest position at time t.

The Bessel functions $J_p(t)$ of the first kind of order p (with p an integer) are similarly related to vibrations of a drumhead with more complicated initial conditions.

Lab 30: Tool Instructions

Boundary Values: Eigenvalues Tool

Parameter Sliders

Use the slider to set the value for the parameter λ.

Press the mouse down on the slider knob and drag the mouse back and forth, or click the mouse in the slider channel at the desired value for the parameter.

Clicking in the plane while a trajectory is being drawn will start a new trajectory.

Buttons

Click the mouse on the [Clear] button to clear all the trajectories from the graph.

Click the mouse on the [Clear to last] button to clear all trajectories except the last one drawn.

Click the mouse on the [Find nearest eigenvalues] button to find the nearest eigenvalues.

Click the mouse on the [y(1)=0] or [y'(1)=0] button to choose the right boundary.

Boundary Values: Solvability and Uniqueness Tool

Parameter Sliders

Use the slider to set the value for k and $y'(0)$.

Press the mouse down on the slider knob for the parameter you want to change and drag the mouse back and forth, or click the mouse in the slider channel at the desired value for the parameter.

Clicking in the plane while a trajectory is being drawn will start a new trajectory.

Buttons

Click the mouse on the [Clear] button to clear all the trajectories from the graph.

Click the mouse on the [Clear to last] button to clear all trajectories except the last one drawn.

Click the mouse on the [Find nearest solution] button to locate the nearest solution.

Bessel's Function: 1st Kind

Parameter Slider

Use the slider to set p, the order of the function.

Press the mouse down on the slider knob for the parameter you want to change and drag the mouse back and forth, or click the mouse in the slider channel at the desired value for the parameter.

Bessel's Equation Tool

Parameter Sliders

Use the slider to set the value for the parameter λ.

Press the mouse down on the slider knob and drag the mouse back and forth, or click the mouse in the slider channel at the desired value for the parameter.

Clicking in the plane while a trajectory is being drawn will start a new trajectory.

Buttons

Click the mouse on the [Clear] button to clear all the trajectories from the graph.

Click the mouse on the [Clear to last] button to clear all trajectories except the last one drawn.

Click the mouse on the [Find nearest eigenvalues] button to find the nearest eigenvalues.

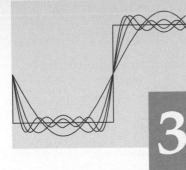

31 Fourier Series

Tools Used in Lab 31

Fourier Series: Square Wave
Fourier Series: Triangle Wave
Fourier Series: Gibbs Effect
Fourier Series: Coefficients

How does a Fourier series converge to a function? What is the difference between a Fourier series and a Maclaurin series? Why is the convergence bad near a discontinuity of the function?

1. Introduction

In 1807, Joseph Fourier announced that an "arbitrary" function $f(t)$ can be represented in terms of sine and cosine functions:

$$f(t) = \sum_{n=0}^{\infty} a_n \cos nt + b_n \sin nt$$

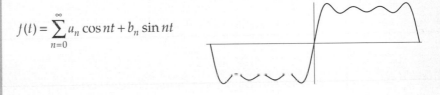

In 1822, Fourier published many examples of such representations. Prior to Fourier, piecewise-defined functions like the square wave and triangle wave were not considered to be "real" functions. The effort to clarify and prove Fourier's claims led to the modern idea of function (which is familiar to all calculus students).

2. Square Wave

Use the **Fourier Series: Square Wave** tool to answer the following questions on approximating a square wave with a Fourier series.

2.1 How does the approximation using five terms of the Fourier series differ from the approximation using four terms of the Fourier series? If you have trouble with this question, look at the approximation using one term, the approximation using two terms, and the approximation using three terms.

2.2 Is the approximation using five terms of the Fourier series better than the approximation using four terms at every point? Discuss.

2.3 In what sense is the approximation using five terms better than the approximation using four terms?

2.4 In general, if one uses n terms of the Fourier series, how many waves (relative maxima) will there be between 0 and π? Relate your answer to the nth term of the series.

2.5 Does the maximum error decrease as the number of terms in the Fourier series approximation increases? Discuss. (The phenomenon that occurs here is called the **Gibbs effect** and occurs near any discontinuity of a function being approximated by a Fourier series.)

2.6 Does the square wave have a Maclaurin series expansion? Discuss.

2.7 For what functions should one try to use Fourier series approximations and for what functions should one use Taylor series?

3. Triangle Wave

Use the **Fourier Series: Triangle Wave** tool to answer the following questions.

3.1 Is the approximation using five terms of the Fourier series better than the approximation using four terms at every point? Discuss.

3.2 In what sense is the approximation using five terms better than the approximation using four terms?

3.3 Does the maximum error decrease as the number of terms in the Fourier series approximation increases? Why does the Gibbs effect not occur? (See Section 4.)

3.4   Does the triangle wave have a Maclaurin series expansion? Discuss.

4. Additional Exercises

4.1   By looking at the graphs of the square wave and the triangle wave, convince yourself that the square wave function is the derivative of the triangle wave function. Explain in words.

4.2   Verify that the Fourier series for the square wave can be obtained by differentiating the Fourier series for the triangle wave term by term.

5. Gibbs Effect

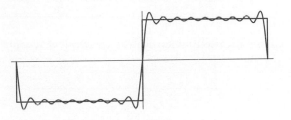

The Gibbs effect is the spike that occurs in a Fourier series approximation near a discontinuity of a function. Use the **Fourier Series: Gibbs Effect** tool and the fact that the square wave is the derivative of the triangle wave (see Exercises **4.1** and **4.2**) to convince yourself that the Gibbs effect has to occur and to answer the following questions.

5.1   Does the maximum error in the Fourier series approximation of the square wave go to zero or do you think it remains larger than some positive value? If so, estimate that value.

5.2   How does the tip of the triangle wave (which occurs at $t = \pi$) manifest itself in the square wave and why?

5.3   By examining the Fourier series approximation of the triangle wave near π, explain why the Gibbs effect has to occur in the square wave.

6. Arbitrary Fourier Series

The **Fourier Series: Coefficients** tool can be used to explore how the coefficients in a Fourier series are related to properties of the function. Good functions to play with are square waves, triangle waves, and sawtooth waves of various periods. Imagine a function shape and try to choose the coefficients, using the sliders to approximate the functions.

Even functions have Fourier series involving only the constant term and the cosine terms. Odd functions have Fourier series involving only the sine terms. An arbitrary function can be expressed as the sum of an even and an old function.

6.1   Show that if $f(t)$ is an arbitrary function then $g(t) = \dfrac{f(t) + f(-t)}{2}$ is even, $h(t) = \dfrac{f(t) - f(-t)}{2}$ is odd, and $f(t) = g(t) + h(t)$.

6.2   What distinguishes functions whose Fourier series involves only terms of the form $\sin nt$ with n even? with n odd?

6.3   What distinguishes functions whose Fourier series involves only terms of the form $\cos nt$ with n even? with n odd?

Lab 31: Tool Instructions

Fourier Series: Square Wave Tool

Buttons

Click the mouse on a button in the upper group to view the graph of a partial series sum.
Click the mouse on a button in the middle group to view overlays of partial series sums.
Click the mouse on a button in the lower group to view graphs of individual series terms.

Fourier Series: Triangle Wave Tool

Buttons

Click the mouse on a button in the upper group to view the graph of a partial series sum.
Click the mouse on a button in the middle group to view overlays of partial series sums.
Click the mouse on a button in the lower group to view graphs of individual series terms.

Fourier Series: Gibbs Effect Tool

Buttons

Click the mouse on a button to view the graphs for a partial series sum.
The upper graphs show the triangle wave, with magnification on the left to see Gibbs effect.
The lower graphs show the square wave, derivative of the triangle wave, with the magnification on the left to see Gibbs effect.

Fourier Series: Coefficients Tool

Parameters Sliders

Use the sliders to change the values for the parameters k and c.
Press the mouse down on the slider knob for the parameter you want to change and drag the mouse back and forth, or click the mouse in the slider channel at the desired value for the parameter.

Buttons

Click the mouse on the [Cosine Series] or [Sine Series] buttons to choose a series.
Click the mouse on the lower set of buttons to select which terms to display: all, odd, or even.

Appendix

IDE Tool Worksheets

Airy's Equation
Airy's Series: Cosine
Airy's Series: Sine
Bessel Equation
Bessel Function: 1st Kind
Boundary Values: Eigenvalues
Boundary Values: Solvability and Uniqueness
Chebyshev's Equation
Chemical Oscillator
Competitive Exclusion
Critical Damping
Damped Forced Vibrations
Damped Vibrations
Damped Vibrations: Energy
Duffing Oscillator
The Eigen Engine
Falling Bodies
Forced Damped Pendulum
Forced Damped Pendulum: Nine Sections
Forced Damped Pendulum: Poincaré Section
Four Animation Paths
Fourier Series: Coefficients
Fourier Series: Gibbs Effect
Fourier Series: Square Wave
Fourier Series: Triangle Wave
The Glider
Golf
Growth and Decay
Hudson Bay Data (Hare-Lynx)
Imperfect Bifurcation
Isoclines
Isoclines as Fences
Laplace: Convolution Example
Laplace: Convolution Theorem
Laplace: Definition
Laplace: Delta Function
Laplace: Derivative
Laplace: Shift and Step
Laplace: Shifting Theorem
Laplace: Solver
Laplace: Transformer
Laplace: Translation
Laplace: Vibrations and Poles
Lissajous Figures
Logistic Growth
Logistic Phase Line
Logistic with Harvest
Lorenz Equation: Discovery 1963
Lorenz Equation: Parameter Grid $0 \leq r \leq 30$

Lorenz Equation: Phase Plane $0 \leq r \leq 30$
Lorenz Equation: Phase Plane $0 \leq r \leq 320$
Lorenz Equation: Sensitive Dependence
Lorenz Equation: Zmax Map
Lotka-Volterra
Lotka-Volterra with Harvest
Maclaurin Series: e^t
Maclaurin Series: Cosine
Maclaurin Series: Sine
Maclaurin Series: $1/\sqrt{1+t}$
Mass and Spring
Matrix Element Input
The Matrix Machine
Newton's Law of Cooling: Cooling Rate
Newton's Law of Cooling: Curve Fitting
Numerical Methods
Numerical Methods: Stepsize Scaling
Orthogonal Trajectories
Parameter Path Animation
Parameter Plane Input
Parametric to Cartesian
Pendulums
Phase Plane Drawing
Pitchfork Bifurcation: Subcritical
Pitchfork Bifurcation: Supercritical
Romeo and Juliet
Saddle-Node Bifurcation
Series Circuits
Simple Harmonic Oscillator
Slope Fields
Solutions
Spruce Budworm: Cusp
Spruce Budworm: Hysteresis
Spruce Budworm: kr-Plane
Spruce Budworm: rx-Plane
Spruce Budworm: Time Series
Sure-Fire Target
Targets
Time Steps
Transcritical Bifurcation
Two Dimensional Equations
2-D Saddle-Node Bifurcation
Uniqueness
Van der Pol Circuit
Vector Fields
Vibrations: Amplitude Response
Vibrations: Input/Output
Vibrations: Phase Response

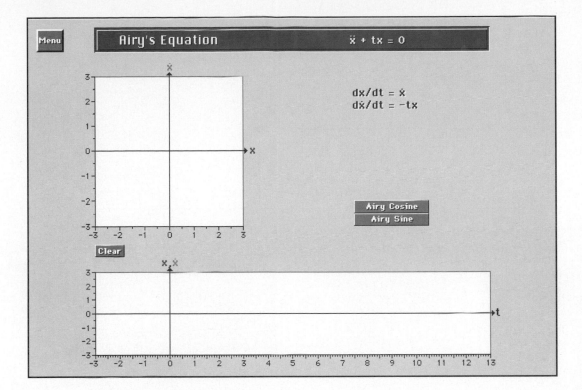

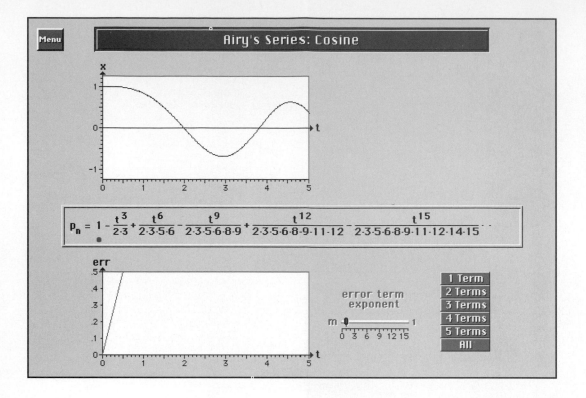

$$p_n = 1 - \frac{t^3}{2 \cdot 3} + \frac{t^6}{2 \cdot 3 \cdot 5 \cdot 6} - \frac{t^9}{2 \cdot 3 \cdot 5 \cdot 6 \cdot 8 \cdot 9} + \frac{t^{12}}{2 \cdot 3 \cdot 5 \cdot 6 \cdot 8 \cdot 9 \cdot 11 \cdot 12} - \frac{t^{15}}{2 \cdot 3 \cdot 5 \cdot 6 \cdot 8 \cdot 9 \cdot 11 \cdot 12 \cdot 14 \cdot 15} \cdots$$

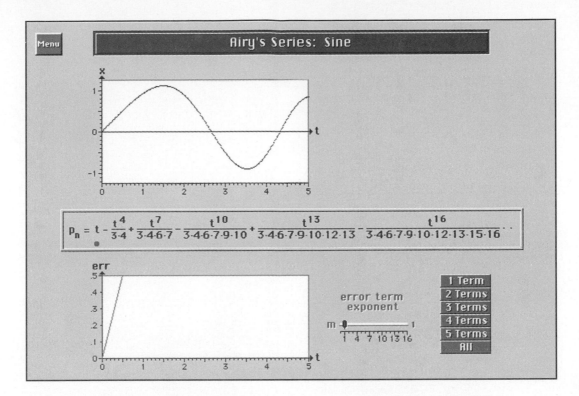

$$p_n = t - \frac{t^4}{3\cdot4} + \frac{t^7}{3\cdot4\cdot6\cdot7} - \frac{t^{10}}{3\cdot4\cdot6\cdot7\cdot9\cdot10} + \frac{t^{13}}{3\cdot4\cdot6\cdot7\cdot9\cdot10\cdot12\cdot13} - \frac{t^{16}}{3\cdot4\cdot6\cdot7\cdot9\cdot10\cdot12\cdot13\cdot15\cdot16} \cdots$$

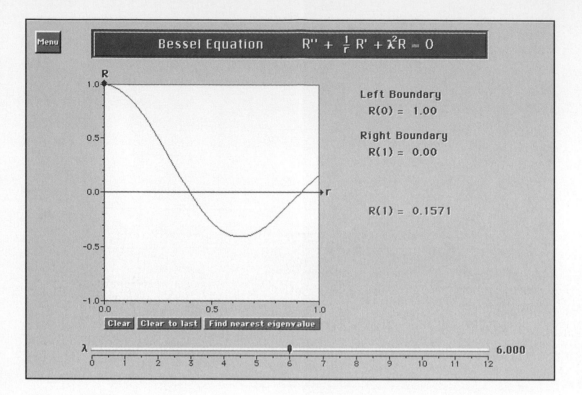

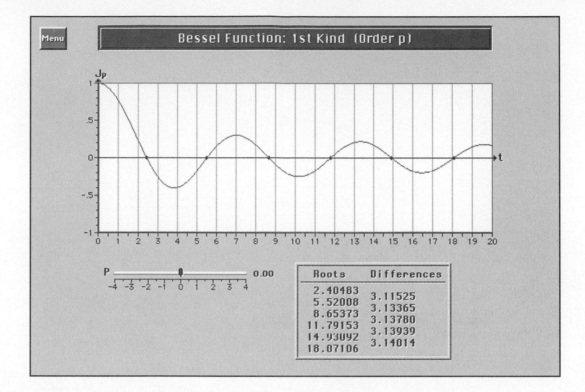

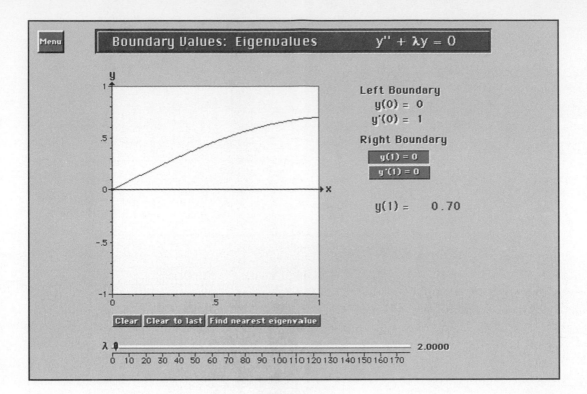

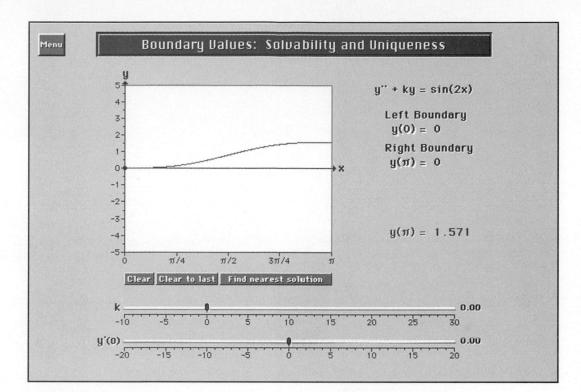

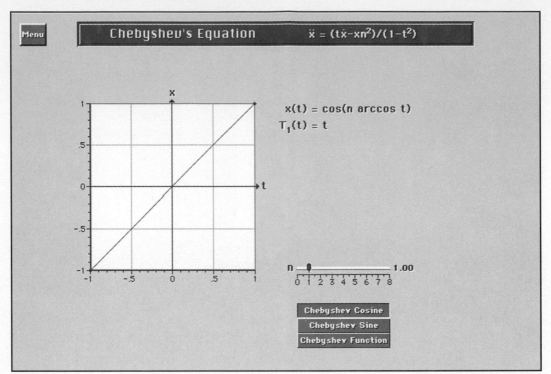

$$\ddot{x} = (t\dot{x} - xn^2)/(1 - t^2)$$

$$x(t) = \cos(n \arccos t)$$

$$T_1(t) = t$$

Chebyshev Cosine

Chebyshev Sine

Chebyshev Function

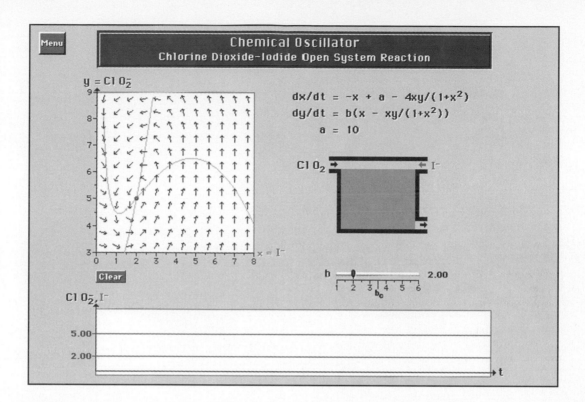

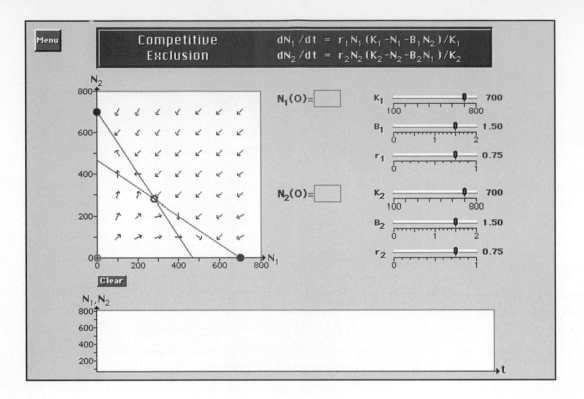

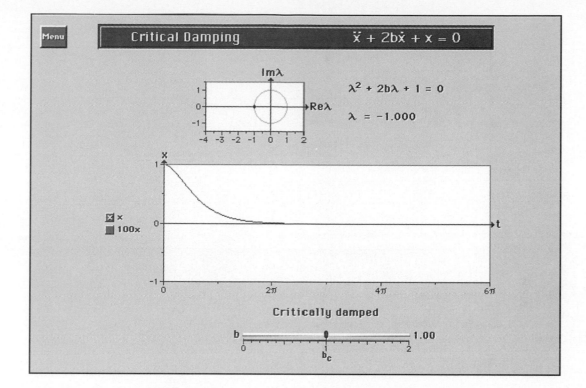

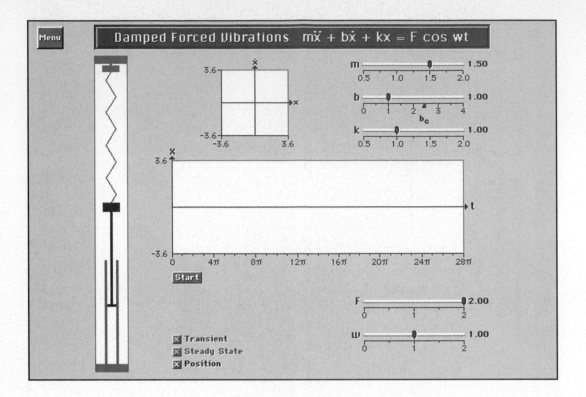

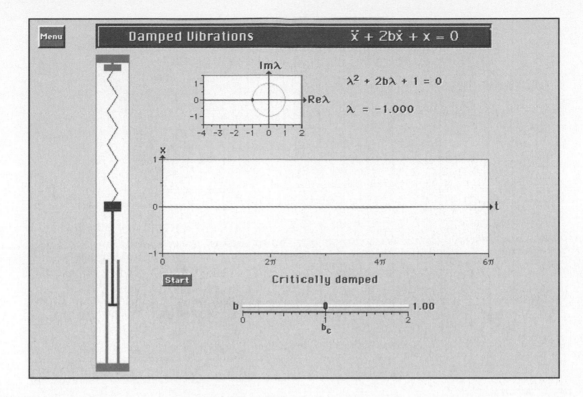

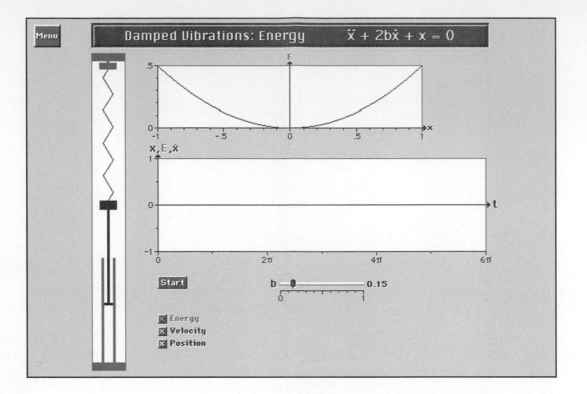

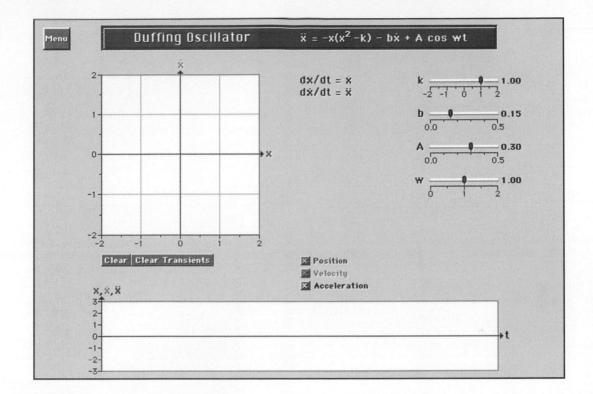

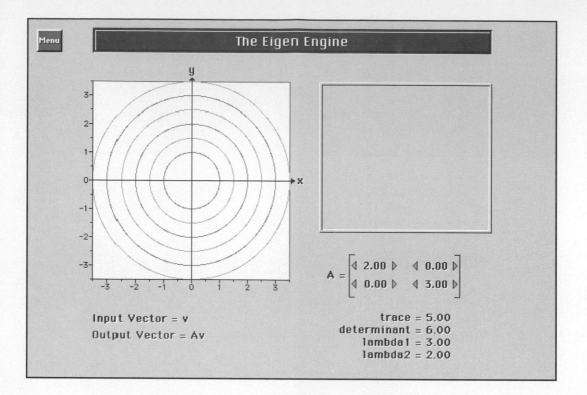

The Eigen Engine

Input Vector = v
Output Vector = Av

$$A = \begin{bmatrix} \triangleleft\ 2.00\ \triangleright & \triangleleft\ 0.00\ \triangleright \\ \triangleleft\ 0.00\ \triangleright & \triangleleft\ 3.00\ \triangleright \end{bmatrix}$$

trace = 5.00
determinant = 6.00
lambda1 = 3.00
lambda2 = 2.00

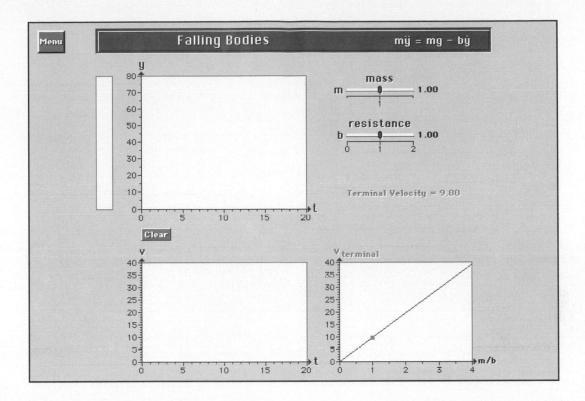

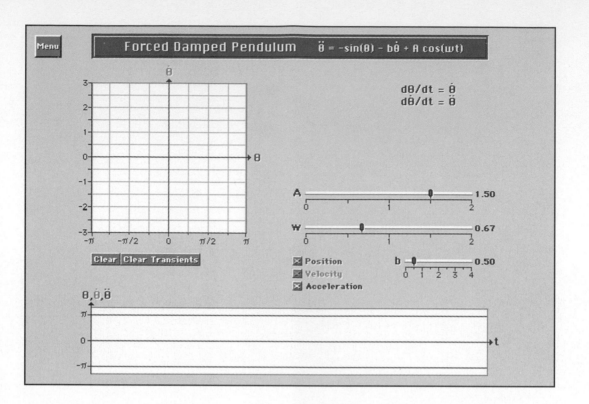

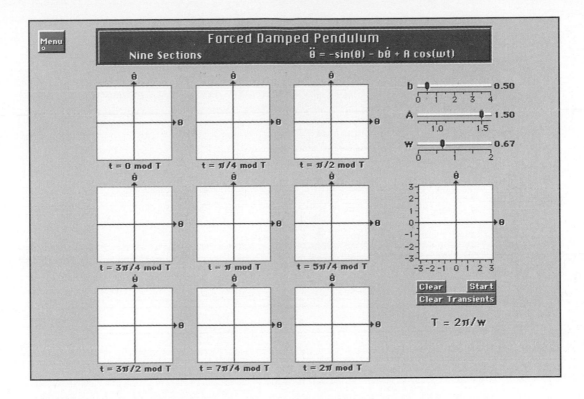

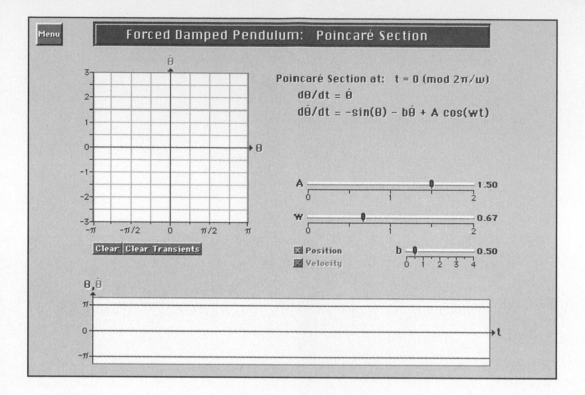

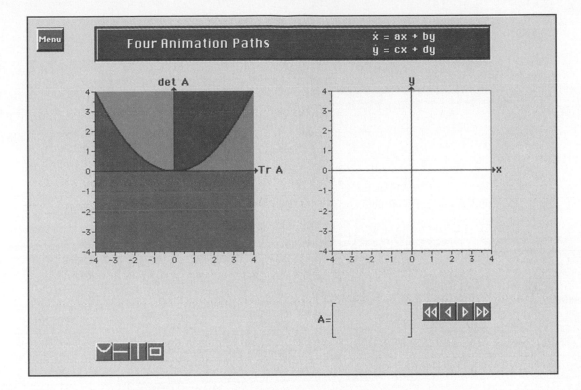

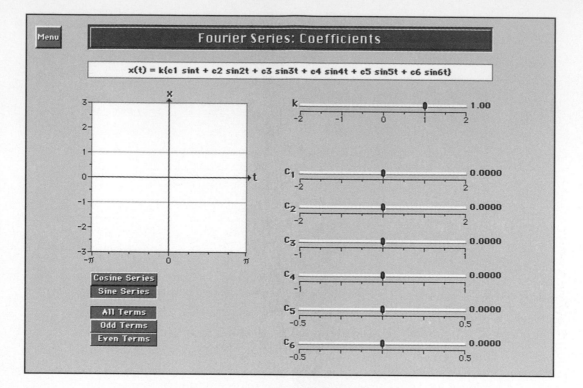

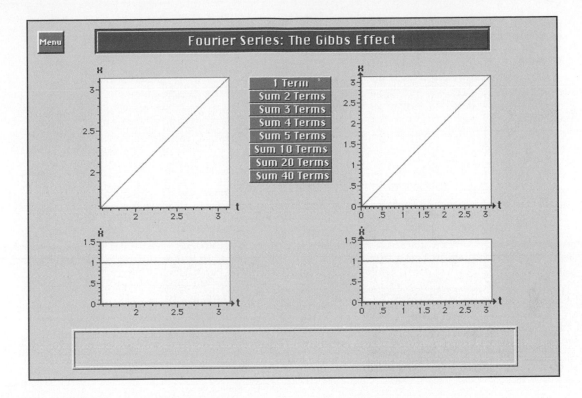

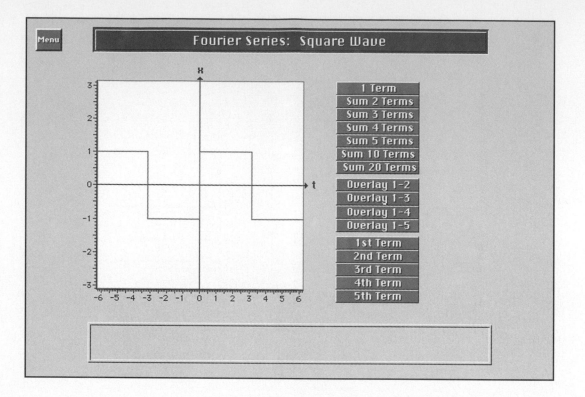

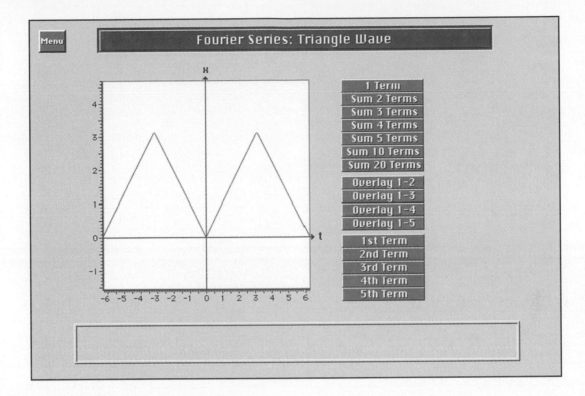

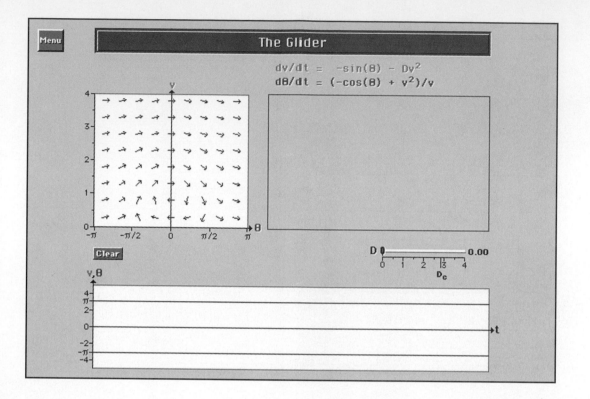

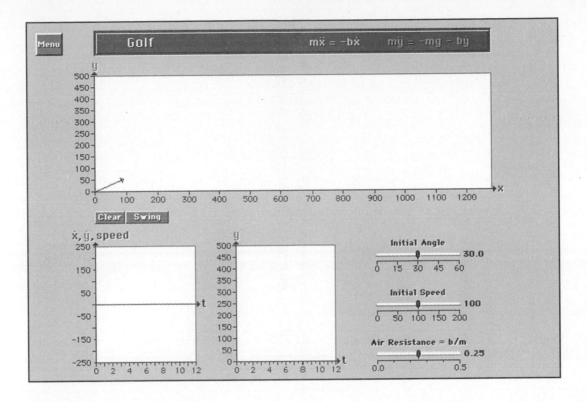

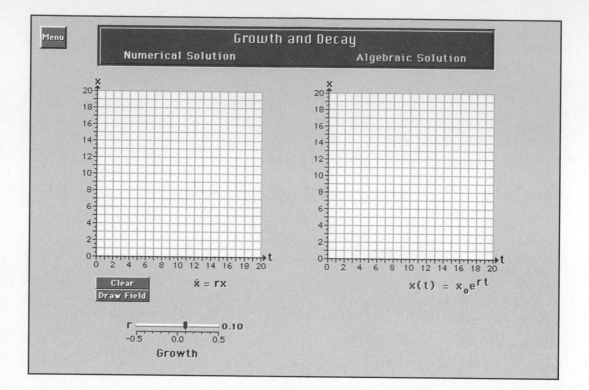

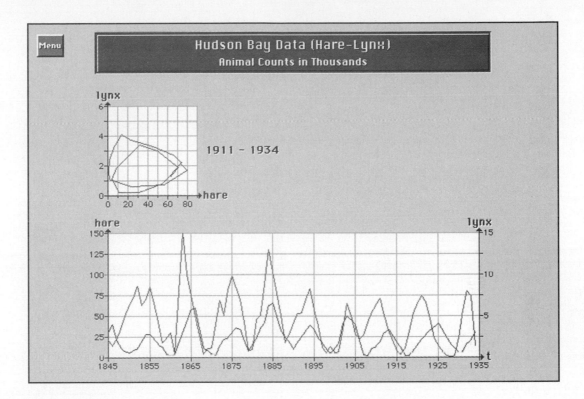

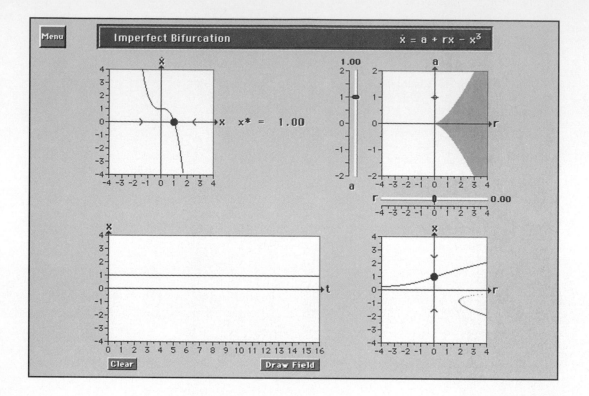

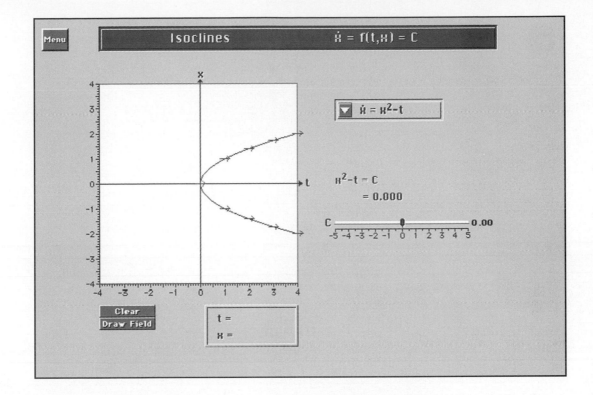

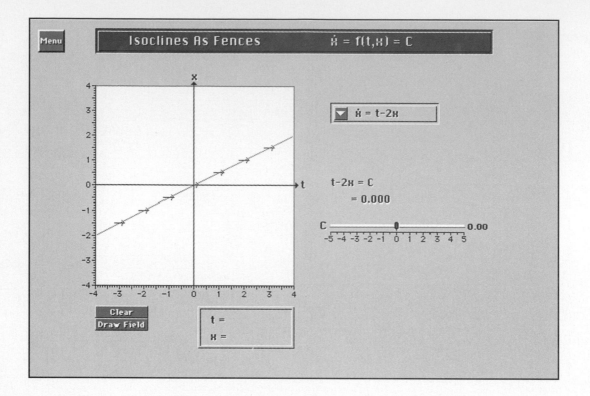

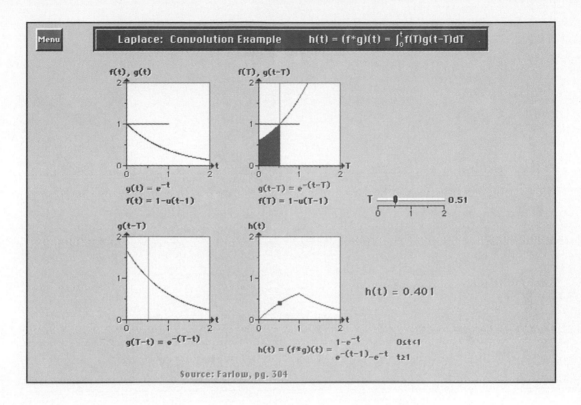

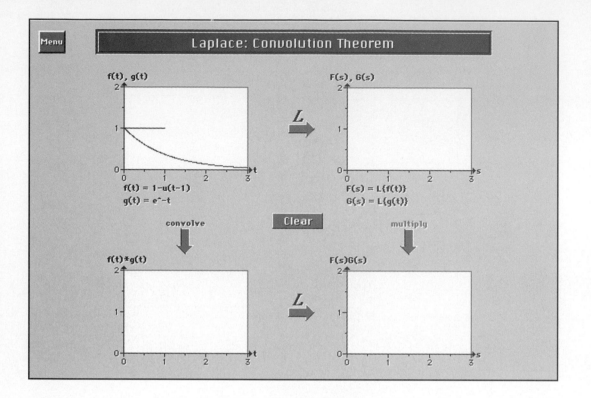

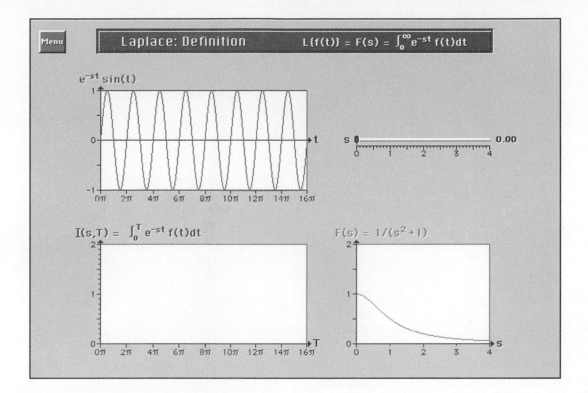

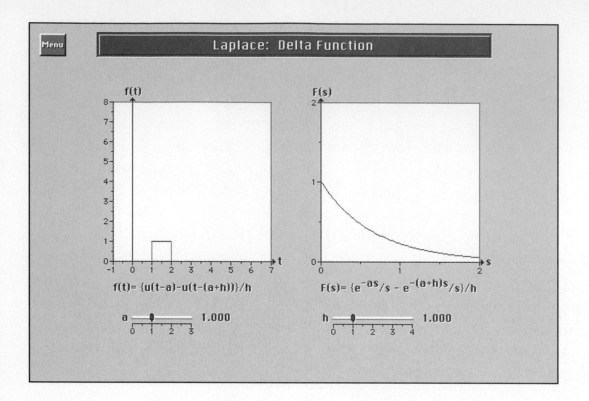

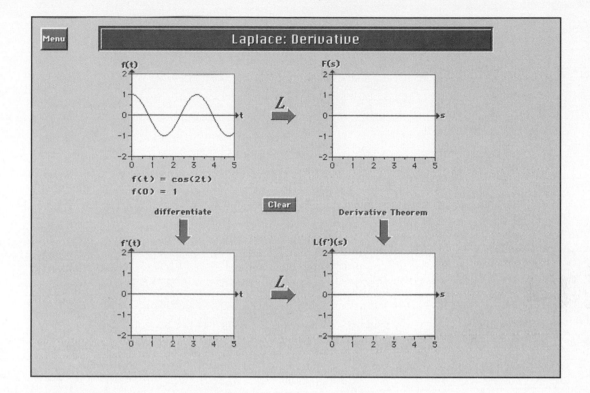

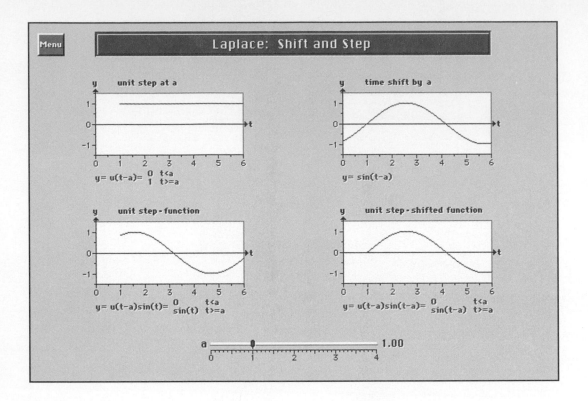

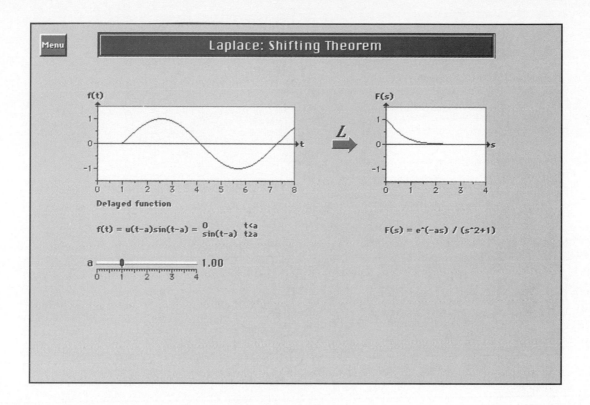

Laplace: Shifting Theorem

$f(t)$

Delayed function

$$f(t) = u(t-a)\sin(t-a) = \begin{array}{ll} 0 & t<a \\ \sin(t-a) & t\geq a \end{array}$$

$$F(s) = e^{-as} / (s^2+1)$$

a ——•—— 1.00

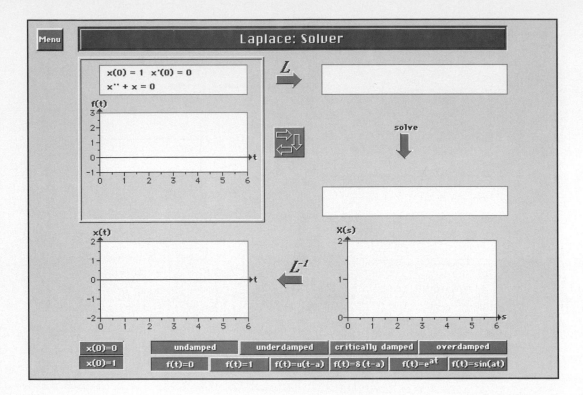

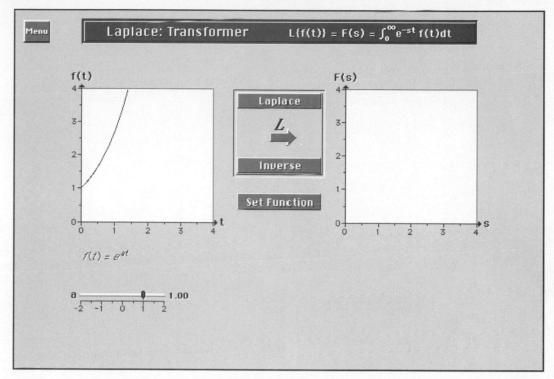

Laplace: Transformer $L\{f(t)\} = F(s) = \int_0^\infty e^{-st} f(t)\,dt$

$f(t) = e^{at}$

a   1.00

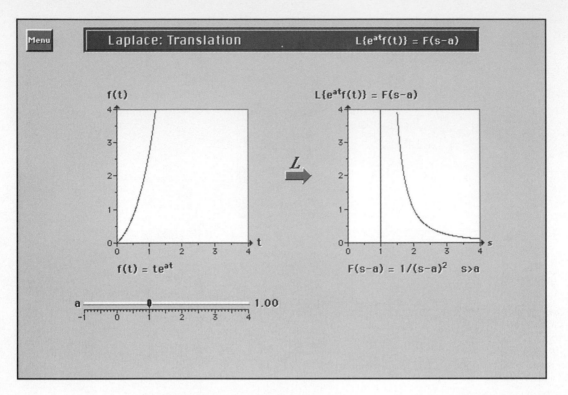

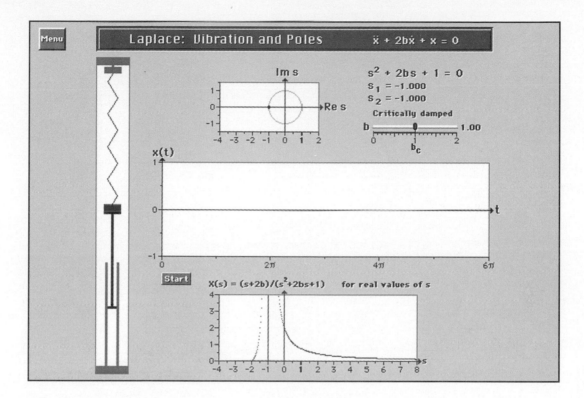

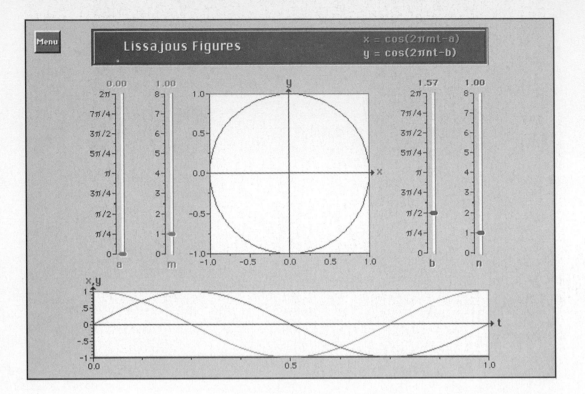

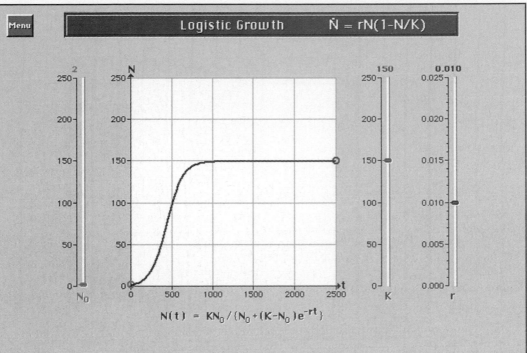

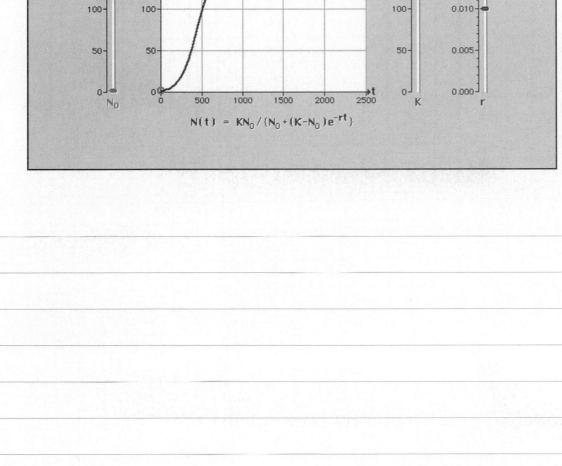

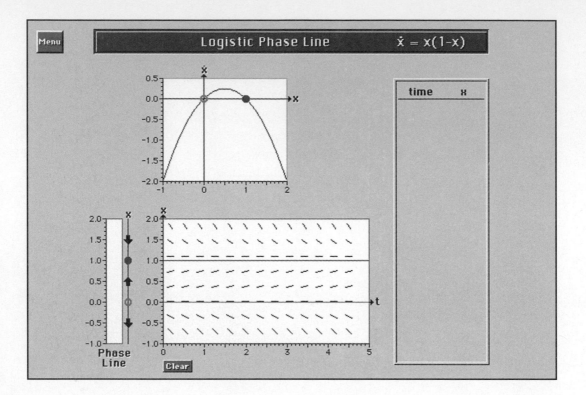

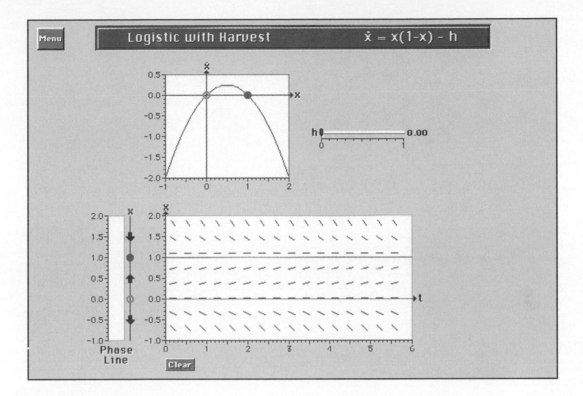

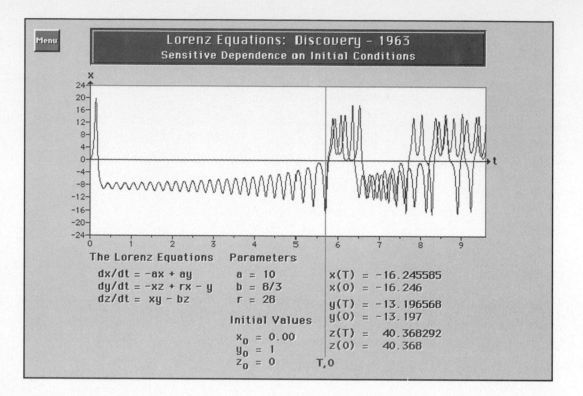

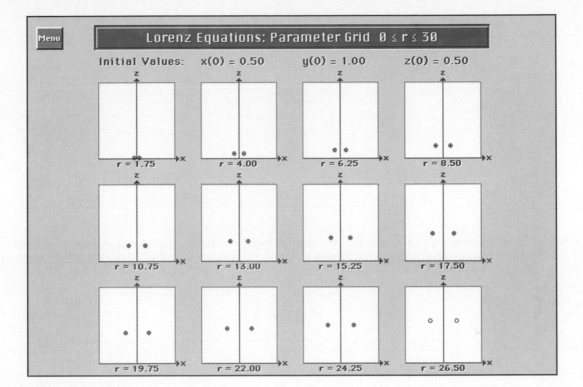

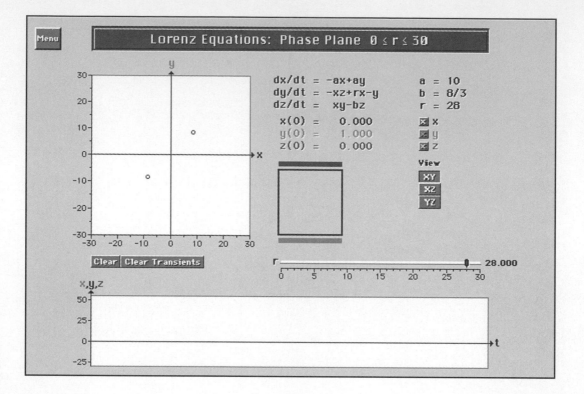

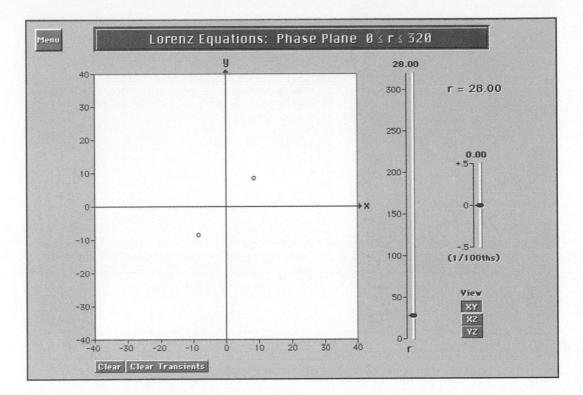

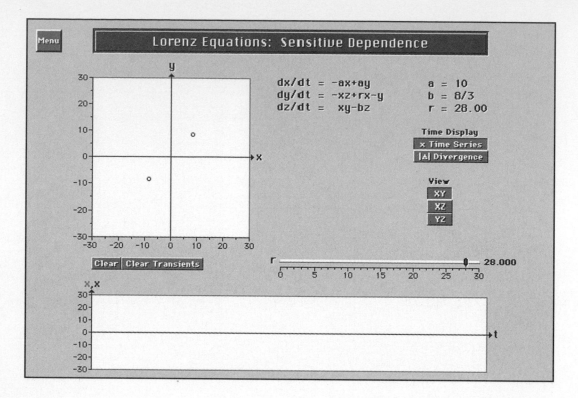

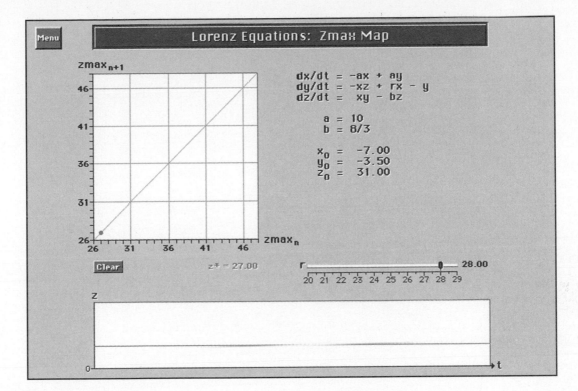

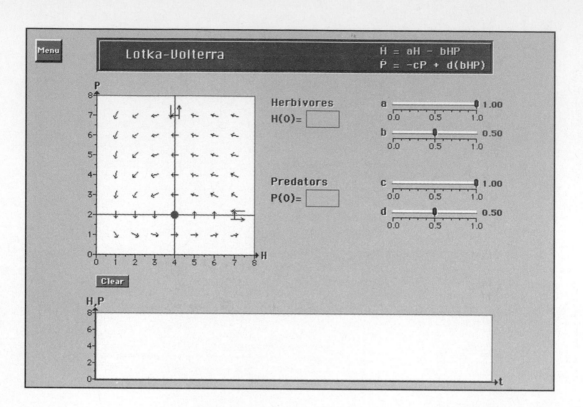

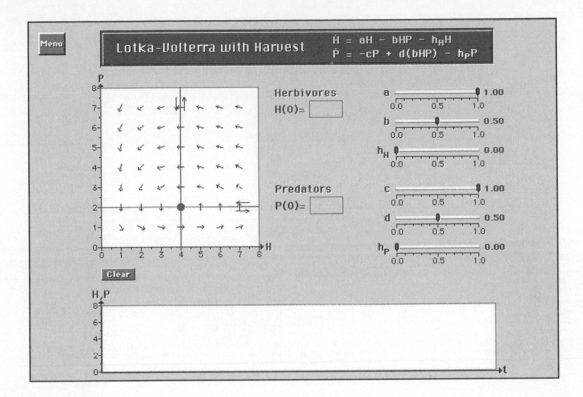

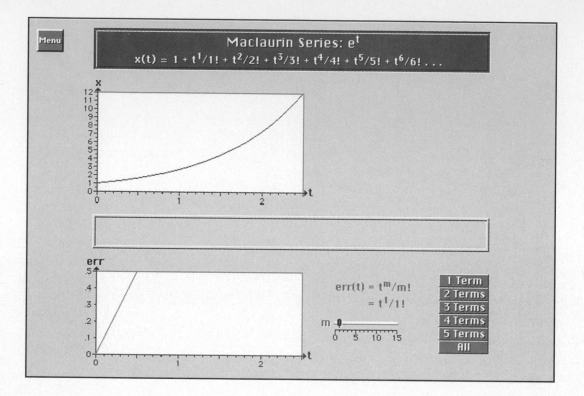

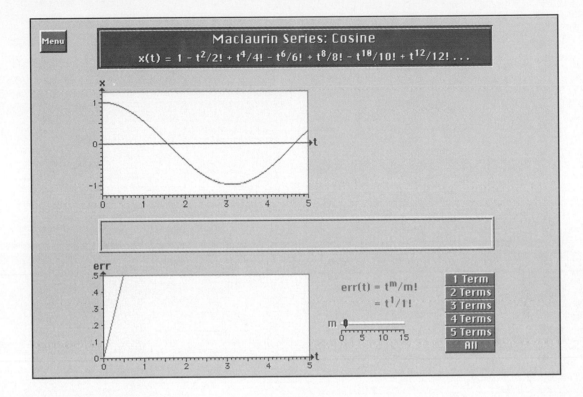

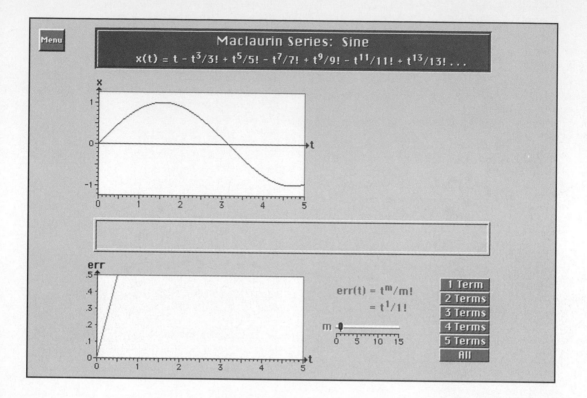

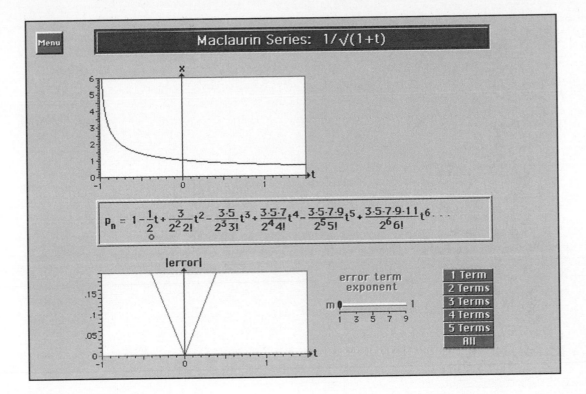

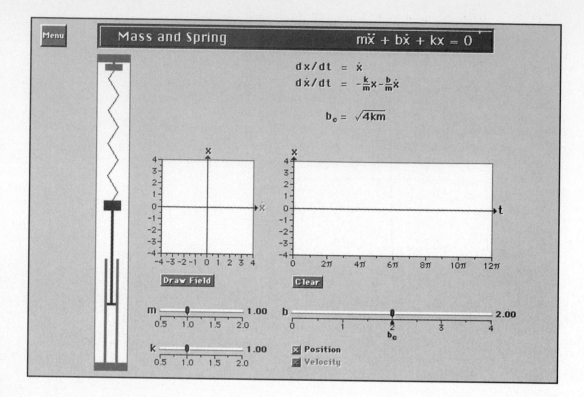

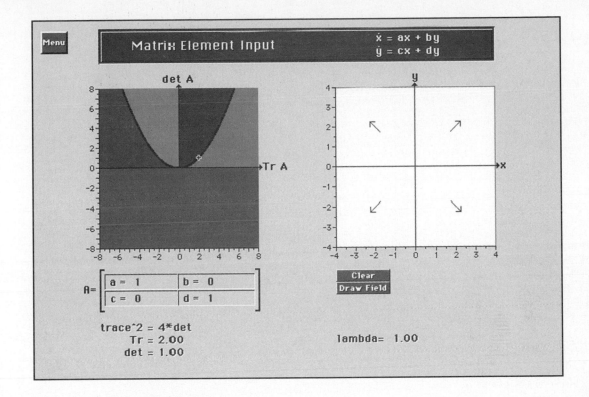

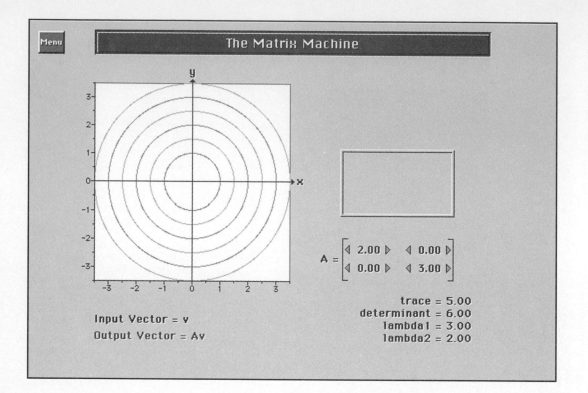

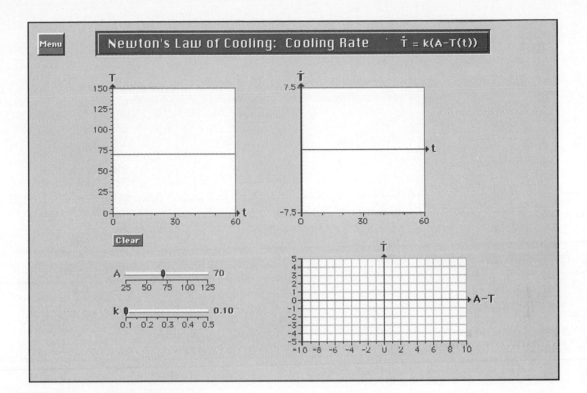

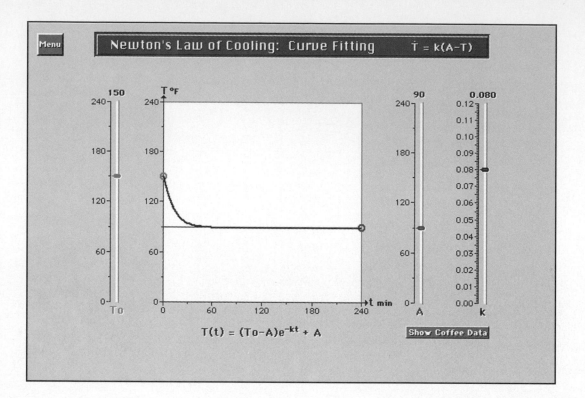

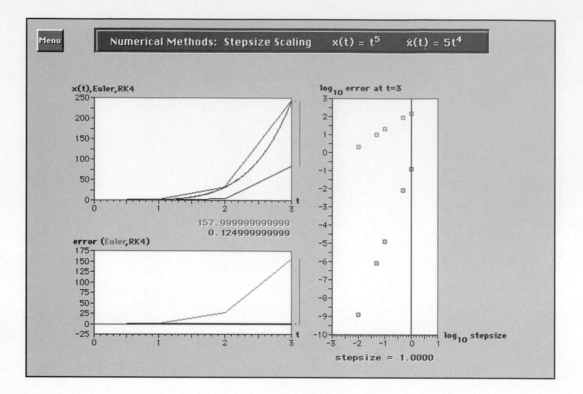

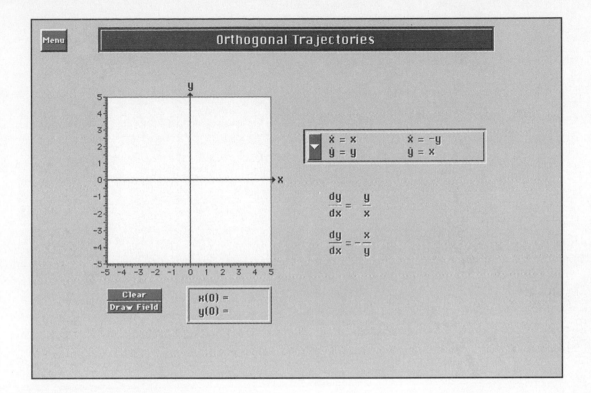

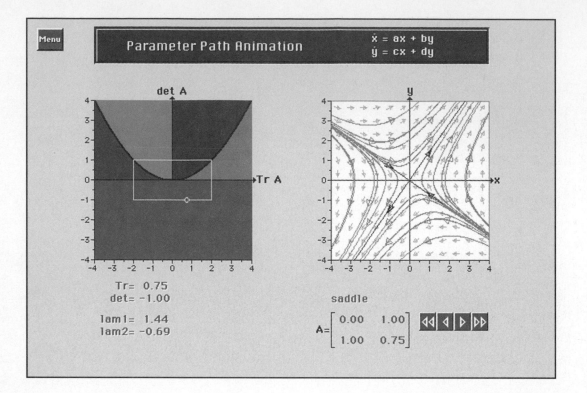

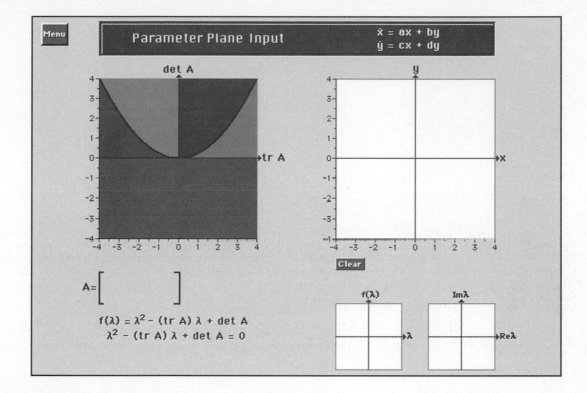

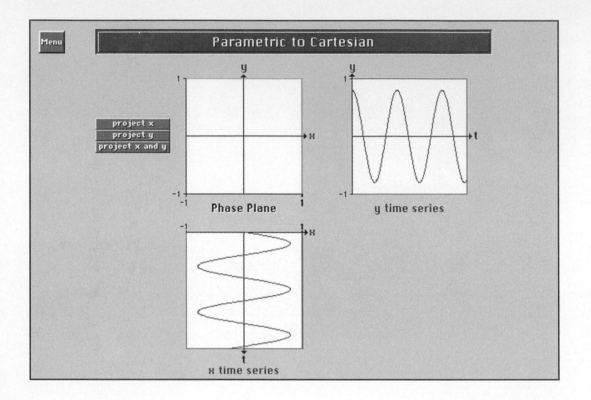

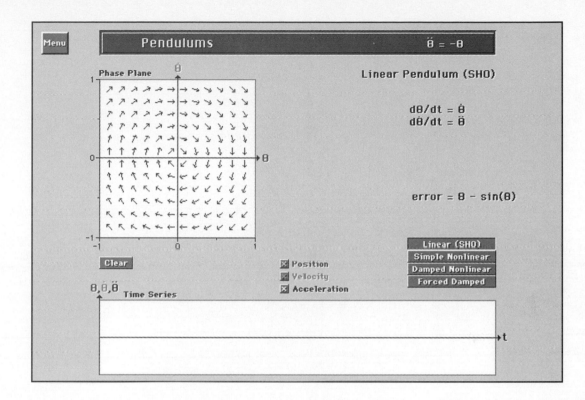

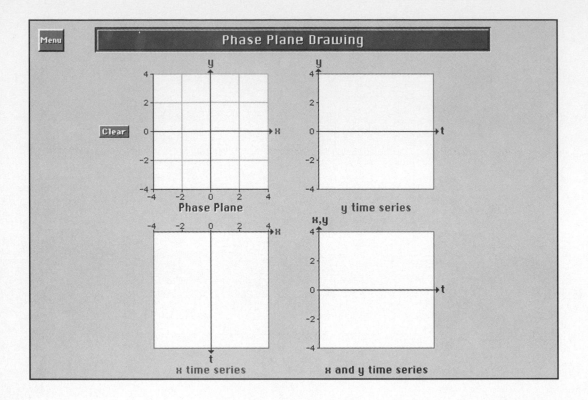

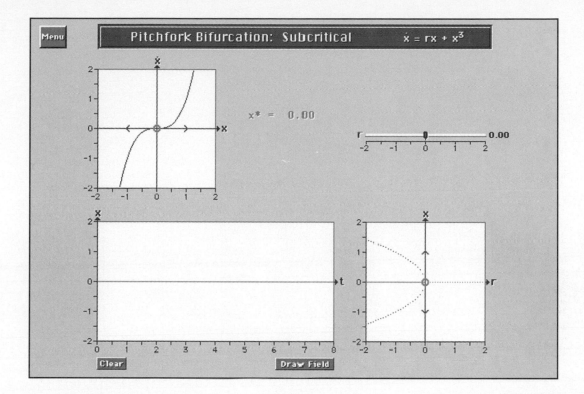

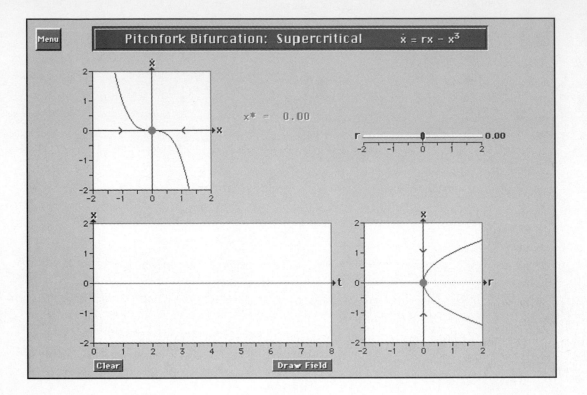

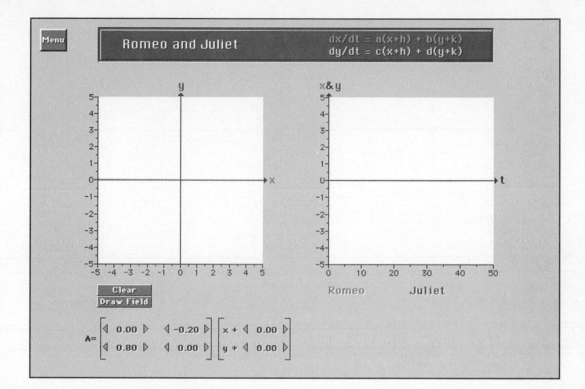

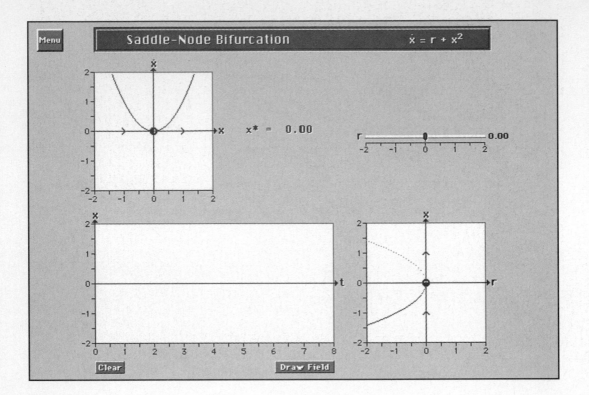

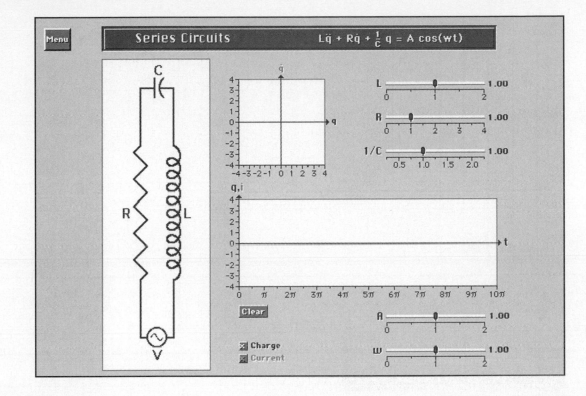

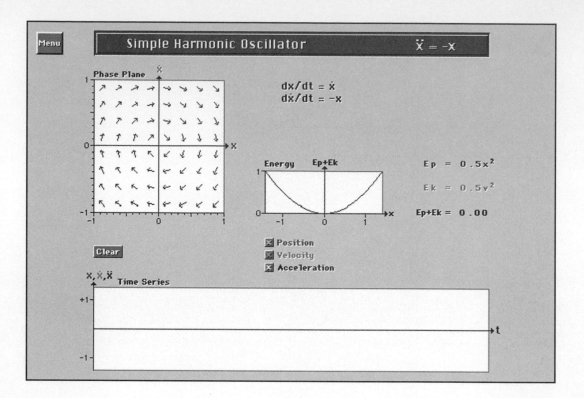

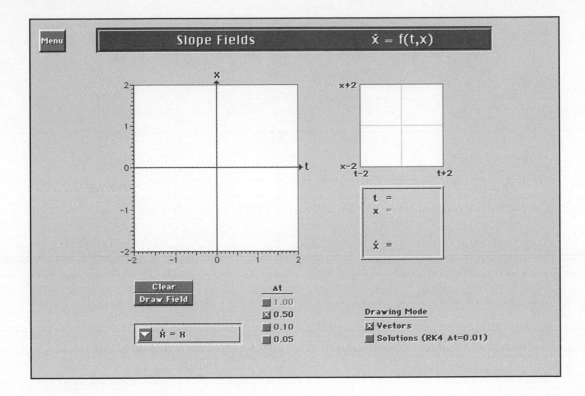

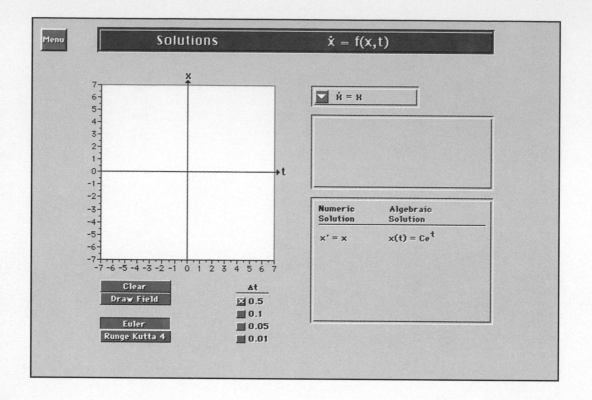

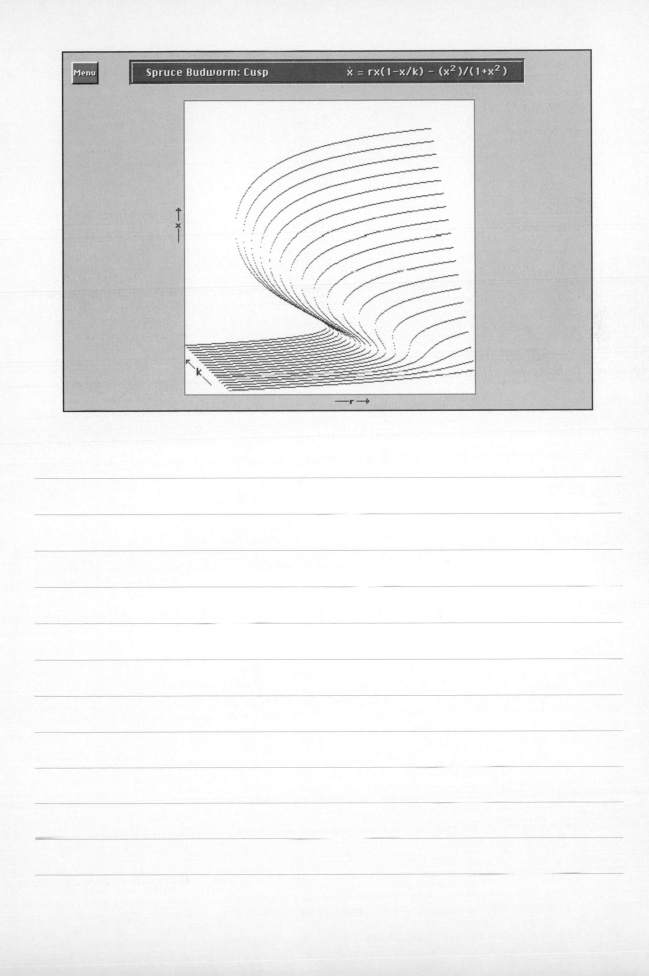

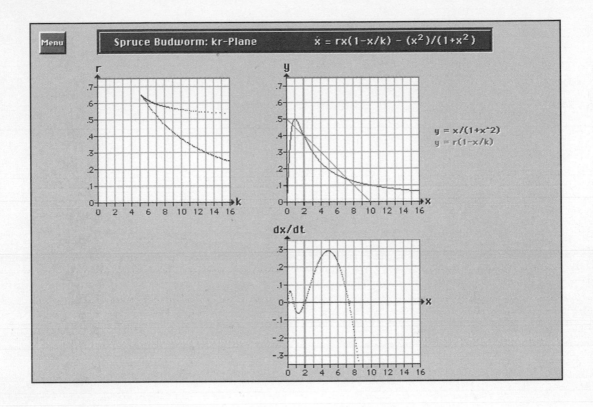

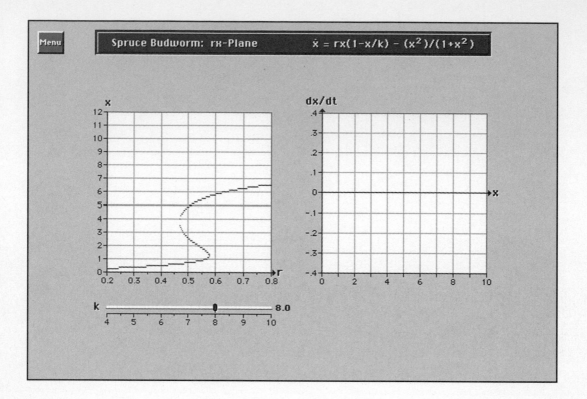

Spruce Budworm: rx-Plane $\dot{x} = rx(1-x/k) - (x^2)/(1+x^2)$

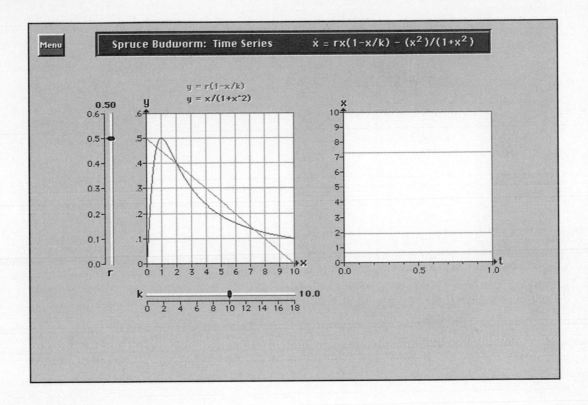

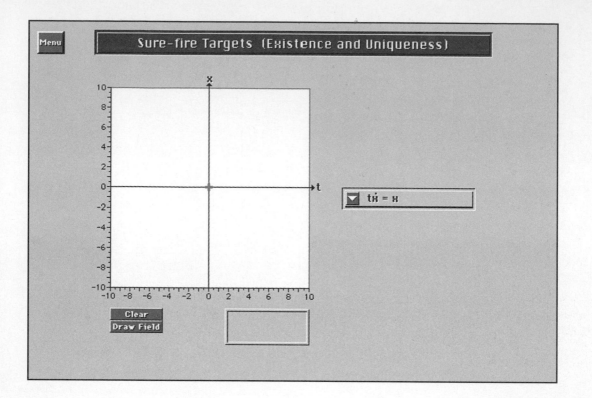

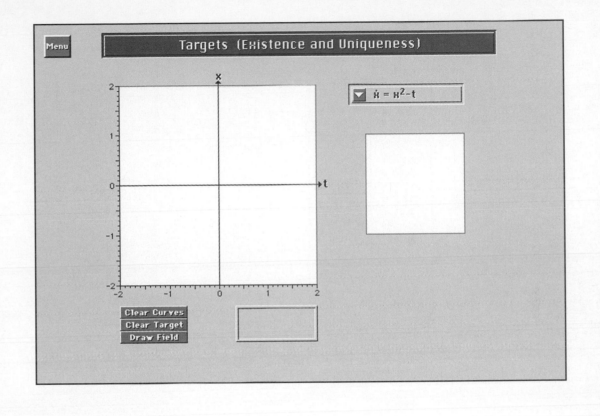

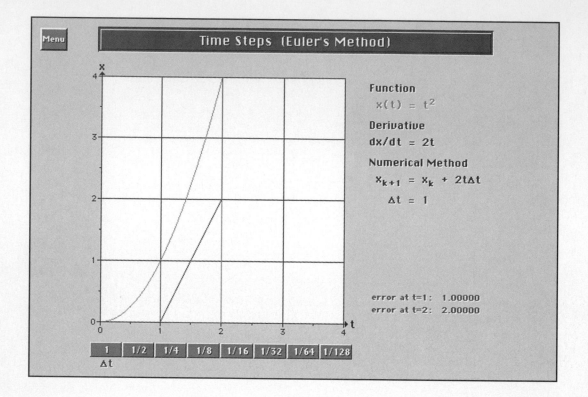

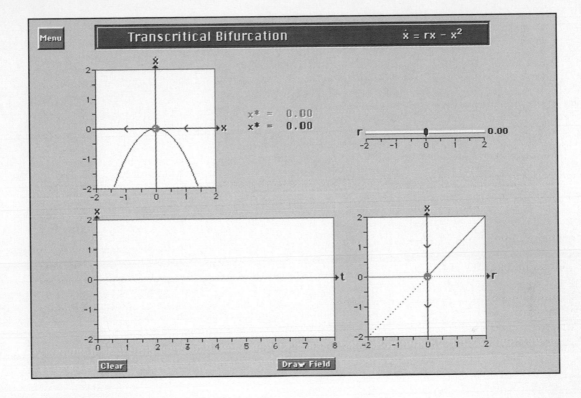

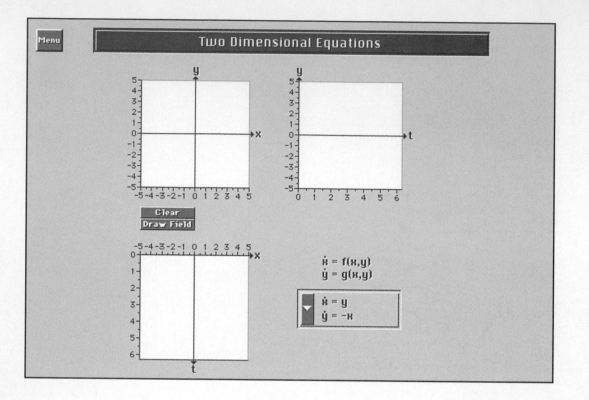

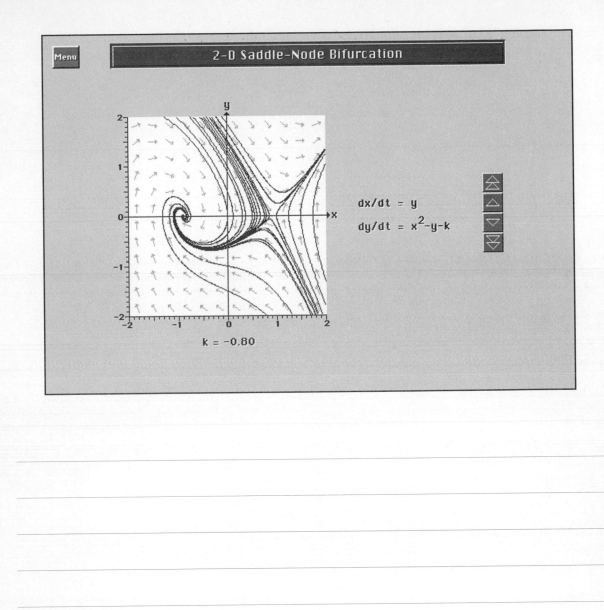

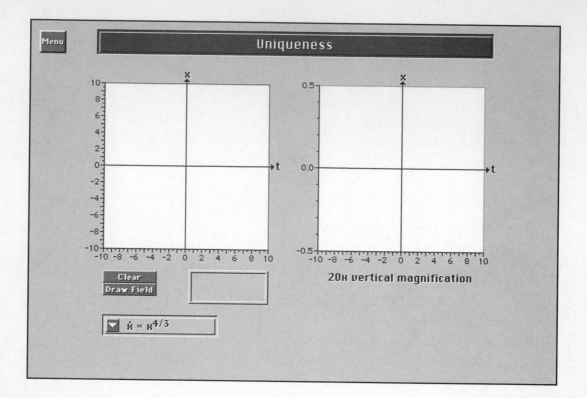

20x vertical magnification

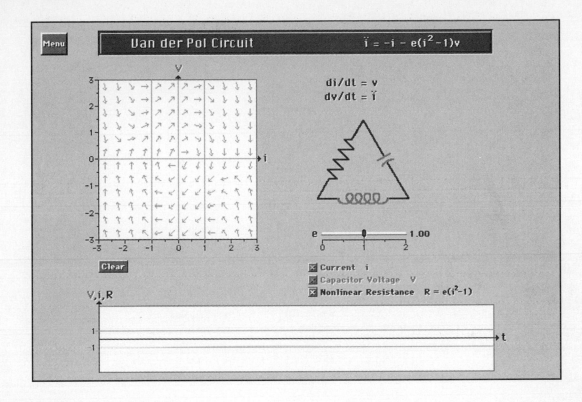

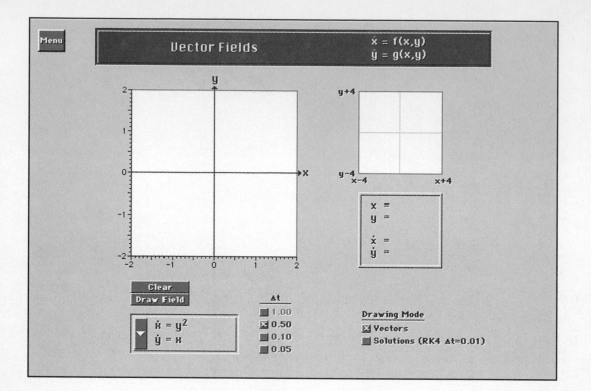

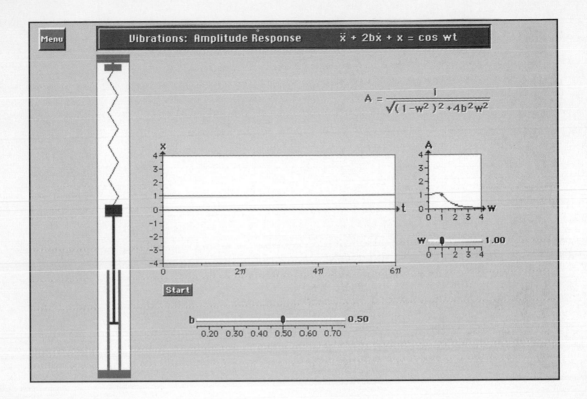

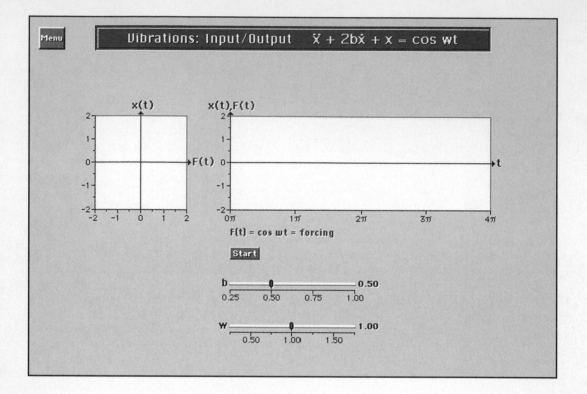

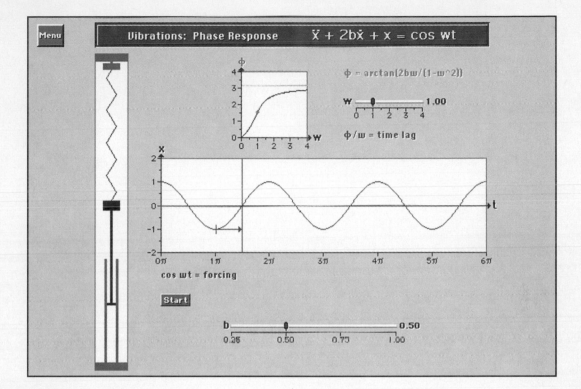

18006776337